S*pringer* **M***onographs* in **M***athematics*

For further volumes:
www.springer.com/series/3733

Jaan Janno • Jüri Engelbrecht

Microstructured Materials: Inverse Problems

Jaan Janno
Institute of Cybernetics
Tallinn University of Technology
Akadeemia tee 21
12618 Tallinn
Estonia
janno@ioc.ee

Jüri Engelbrecht
Institute of Cybernetics
Tallinn University of Technology
Akadeemia tee 21
12618 Tallinn
Estonia
je@ioc.ee

ISSN 1439-7382 Springer Monographs in Mathematics
ISBN 978-3-642-21583-4 e-ISBN 978-3-642-21584-1
DOI 10.1007/978-3-642-21584-1
Springer Heidelberg Dordrecht London New York

Library of Congress Control Number: 2011936963

Mathematics Subject Classification (2010): 35R30, 35L51, 35I77, 74J25, 74J35

Cover design: VTeX UAB, Lithuania

Printed on acid-free paper

Springer is part of Springer Science+Business Media (www.springer.com)

Preface

This book is about the mathematical treatment of inverse problems related to material characterisation. We realised that we have to write a much longer treatise than usual research papers in order to describe our ideas in a proper informative way.

There is always a question about theory and practice like the iconic "the chicken and the egg" causality dilemma. We might say that an apple fell first and then Isaac Newton presented a theory. On the other hand, the ideas of Paul Dirac are excellent examples of how theory precedes practice. And certainly much has been written about the balance of theory and practice. We hope that, although we are theorists ourselves, we have taken the idea of balance seriously.

Indeed, this way or another, in material science the will and the need to look into materials for determining the physical or geometrical properties or residual stresses goes back to the history. The audible ring of a Damascus sword blade or a church bell was an indication of quality, for example. And farmers could estimate the ripeness of a water-melon by tapping it and listening to the sound. However, contemporary technology and materials need more "advanced" techniques, and the information we are interested in is more sophisticated.

Here we deal with Non-Destructive Evaluation (NDE) of material properties with the focus on microstructured materials. The tool for this is a wave which propagates through a specimen or a structural element. A wave travelling in materials and propagating over a certain distance collects and "encodes" the information on its path. The problem is how to "decode" this information.

For many practical purposes the material is assumed to be homogeneous. In this case the "decoding" is rather simple—the flight time of a wave (a signal) permits the sound velocity to be determined and, from that, some information about the material properties (density, modulus of elasticity) can be deduced. However, the contemporary materials are much more complicated and their internal structure, i.e., the microstructure, affects the result. Moreover, there is a need to evaluate the properties of the microstructure. The list of such microstructured materials widely used in modern technology is long: alloys, ceramics and composites, functionally graded materials, granular materials and nano-materials, biomedical and optical materials, etc. Consequently, the methods of the NDE must be based on the adequate analysis of the effects which will be "encoded" by a propagating wave, and the following "decoding" must be properly built up. Following these ideas, the mathematical modelling of waves in microstructured solids must give a well-grounded basis for the analysis. The focus is on dispersion which is the leading effect for waves in such materials. This is exactly the starting point for the book. After proposing (with suitable assumptions) a sound mathematical model, we discuss the number of unknowns to be inversely identified, then establish the uniqueness of a solution and only then propose the ideas for solving the inverse problems. We hope that such a consistent approach will build a proper basis for practical applications.

We have published several research papers on this topic over the last six years (see references). However, the book is not simply a collection of these papers but much is rewritten to cast the material into a unified whole and much is added in order to cement the ideas.

The book could not have been written without the support of the Institute of Cybernetics at Tallinn University of Technology and the Centre for Nonlinear Studies (CENS) of the Institute. The research has been supported by the CENS-CMA project Cooperation of Estonian and Norwegian Scientific Centres within Mathematics and its Applications (Marie Curie Host Fellowship for the Transfer of Knowledge, Contract MTKD-CT-2004-013909), the target funding from the Estonian Ministry of Education and Research (SF Nonlinear waves and stress analysis, SF Nonlinear waves and complexity, SF Mathematical models with nonlinearities, incomplete information and structural complexity) and grants from the Estonian Science Foundation (6018, 7728). We would like to thank our colleagues in CENS and abroad for valuable discussions. We appreciate very much the invaluable assistance of Martin Peters and Ruth Allewelt from Springer-Verlag for producing the book and acknowledge the excellent help from Michael Easthams on style and English grammar of the manuscript. What is most important, our special thanks are to our families who understand us and tolerate our long working hours.

Tallinn, Estonia

Jaan Janno
Jüri Engelbrecht

Contents

Chapter 1
Introduction

There are two notions used in the title of the book: microstructured solids and inverse problems. We start here by giving short descriptions of these notions.

All materials actually have an internal structure down to molecules and atoms. In conventional theories of continua this internal structure is homogenised and the result is the continuum theory of homogeneous materials. These theories have played an important role in deriving the methods of analysis, and many practical applications are based on the theory of homogeneous media. However, beside molecules and atoms, the discreteness of the material may be expressed at larger scales than atomic. Indeed, polycrystalline solids, alloys, ceramic composites, functionally graded materials, biological tissues etc. are all characterised by certain internal complex structures which have an intrinsic space-scale. In mechanics, when such materials are under static loading, the models can still be homogenised, i.e., material properties averaged over certain scales. If loading is of high-frequency then the wavelengths become measurable with the intrinsic space-scale of the material and then in analysis the internal structure must be taken into account. So here we take the notion of microstructured solids as a continuum (macrostructure) with an internal structure (microstructure) and use the corresponding mathematical models.

In general terms, by an inverse problem we consider the reconstruction of causes from given consequences. In more detailed terms the inverse problem means the determination of parameters from measured data once the general (mathematical) structure of the model is given.

These two notions together mean that we are interested in determination of physical parameters of microstructured solids. However, this is a very large field of analysis and applications because of the plethora of materials and possible physical effects accompanying wave motion. The excellent description of various methods, techniques and state of art applications of the NDE can be found in the book by Liu and Han [46] and in a Handbook [32].

Our idea is to elaborate here in detail the theory of ultrasound NDE for microstructured solids based on a Mindlin-type model. It has been shown [3, 18, 45] that this model is quite general and describes the dispersive effects caused by a microstructure with a needed accuracy. In addition to that, the model can be easily

J. Janno, J. Engelbrecht, *Microstructured Materials: Inverse Problems*,
Springer Monographs in Mathematics,
DOI 10.1007/978-3-642-21584-1_1,

modified by considering physical nonlinearities on both macro- and micro-scales. In this case, the inverse problems are much more complicated but fortunately nonlinear effects, when balanced by dispersion, give rise to a special type of waves called solitary. The properties of solitary waves can be used for solving the inverse problems and actually such an analysis opens novel doors to nonlinear NDE.

The book is organised as follows. It can be divided into three parts: the first part deals with the general description of inverse problems and mathematical models; the second part is devoted to linear waves, and the third part to nonlinear waves. Finally, a short chapter with closing remarks describes the results in general terms and envisages the further studies.

Chapter 2 deals with general inverse problems and the NDE. Two viewpoints are described separately: a mathematical viewpoint and a practical viewpoint (i.e. realisation). Such an opening is related to our deep understanding that theory and practice must be balanced. One has to understand what is a theoretical model, what are the physical effects described by such a model and what is possible to measure. The uniqueness and stability of solutions of inverse problems indicate the sufficient informativity of the data to be collected from measurements. From a practical viewpoint, the generation of ultrasound wave-fields is briefly described.

In Chap. 3, the governing mathematical model is derived. The importance of basic principles—determinism, equipresence, admissibility—is stressed. The mathematical model is based on the Mindlin [50] theory as represented by Engelbrecht et al. [16, 17] for linear and nonlinear cases. The corresponding one-dimensional systems of equations describe the motion of macro- and microstructure, respectively. Further we refer to these systems as coupled systems. In addition, using a certain asymptotic technique, it is possible to derive the corresponding higher-order single equation which displays clearly the hierarchical nature for a propagating wave in a microstructured material. It means that, depending on the ratio of the internal scale over the wave-length, the influence of the microstructure is either weak or strong. In other words, this indicates the strength of dispersion. In this way, we have two models for the further analysis: (i) a coupled system and (ii) a hierarchical equation. Finally, in this chapter, the parameters characterising these two models are discussed in order to prepare the ground for solving the inverse problems.

Chapters 4 and 5 are devoted to linear waves and the corresponding inverse problems. The main attention in Chap. 4 is focused on the dispersion relations derived for the coupled system and the hierarchical equation. The conditions for normal and anomalous dispersion are established. It is shown that the phase and group velocities may be strongly influenced by dispersive effects which will be used later for solving the inverse problems. The solution for right-propagating waves and their special form—Gaussian wave packets—are presented. This analysis is then used in Chap. 5 for considering the inverse problems. In case of harmonic waves, the coupled system and the hierarchical equation are dealt with separately. The conditions for the uniqueness and the stability of solutions are established. As said before, the corresponding solution of an inverse problem uses the strong dependence of phase velocities on the parameters of the microstructure. In the case of more general linear waves, the spectral decomposition is used to extract the harmonic counterparts and solve the inverse problem. Here two boundary conditions: of deformation-type and of displacement-type, are analysed. In the particular case of the Gaussian wave packet which is important in the practice, phase and group velocities and amplitude changes are used to reconstruct the parameters.

Chapters 6 and 7 deal with nonlinear problems. A short description of solitons and solitary waves in Chap. 6 serves as an introduction. Then the solitary wave solutions are derived for the coupled system and for the hierarchical equation. The main result is that a solitary wave in a microstructured material takes an asymmetric form, while in a homogeneous material a solitary wave is of a symmetric form. The conditions of existence of such asymmetric solitary waves are determined, and it is shown how these conditions depend on the parameters of the microstructured material. The asymmetry of the profile is used in Chap. 7 for solving the corresponding inverse problems. As in Chap. 5, here also the uniqueness (and in a particular case the stability) of solutions is proved and then the ideas for solving inverse problems are presented.

The analysis in Chaps. 4–7 is mainly mathematical containing lemmas and theorems with related proofs. Most important proofs that illuminate the essence of methods are presented in the main text. Other proofs and computations are collected into the last sections of each chapter. For a reader with a general interest in inverse problems the latter sections may be skipped leaving them to mathematicians.

The final Chap. 8 summarises the results together with ideas for possible further studies.

Chapter 2
Inverse Problems and Non-destructive Evaluation

2.1 Inverse Problems from a Mathematical Viewpoint

In deterministic physical processes two types of quantities occur: causes and consequences. The consequences are states of the process (deformation, temperature, voltage etc.) and the causes are the medium or material parameters (elasticity parameters, density, conductivity etc.), initial states or boundary conditions. The problem of determining the consequences from given causes is called the *direct* or *forward problem*, and the reconstruction of causes from given consequences is called the *inverse problem*.

Usually a mathematical model of the process consists of differential, differential-algebraic or integral(-differential) equations. Inverse problems serve for two purposes:

(1) testing the relevance of the mathematical model—solving the inverse problem several times with different data. An approximate coincidence of the solutions proves the relevance of the model, whereas a big difference between the solutions shows the irrelevance;
(2) practical determination of the parameters, initial states, etc. for particular materials and media.

The most important application fields of inverse problems are in geophysics, medical and industrial tomography and materials science. There is a large amount of literature devoted to this topic. Good overviews can be found in monographs by Anger [1], Colton and Kress [7], Gladwell [29], Isakov [33], Kabanikhin and Lorenzi [41], Romanov [56], Santamarina and Fratta [60], Trujillo and Busby [70].

Once the inverse problem is posed, one must answer some basic mathematical questions. One of them is the well-posedness. A problem is called *well-posed* in the sense of Hadamard [30] if the following three requirements are met.

1. The solution exists.
2. The solution is unique.
3. The solution is stable with respect to small errors of the data.

J. Janno, J. Engelbrecht, *Microstructured Materials: Inverse Problems*,
Springer Monographs in Mathematics,
DOI 10.1007/978-3-642-21584-1_2,

If at least one of these requirements is violated, the problem is called *ill-posed.*

In most cases, the existence of the solution is rather a theoretical issue. But the uniqueness and stability are very important in practice.

The uniqueness is related to the question: what is the minimal amount of information necessary for the determination of the solution? Often one can use many measurements in the inverse problem, but the solution is still nonunique. For instance, in Chap. 7 we will show that a single solitary wave does not contain enough information to recover all 5 coefficients of a nonlinear wave equation under consideration, no matter how many measurements are gathered from this wave. Therefore, a mathematical analysis of the uniqueness must be implemented before the practical solution. There are several methods to study the uniqueness in inverse problems [1, 7, 33, 56]. In case the solution to be determined is a vector of functions then it belongs to an infinite-dimensional functional space, and hence methods of functional analysis such as fixed point arguments or monotonicity methods could be applied. In case the solution is a vector of scalars then it is an element of a finite-dimensional space and methods of algebra or real analysis (e.g. mean value theorems) could also be useful. In our book we have the latter case.

In mathematical literature, stability is usually meant in the asymptotical sense: if the data error ε_d approaches zero then the solution error ε_s also approaches zero. When such a stability is violated, one can "improve" the problem by means of the regularisation [21, 68]. This consists in replacement of the original problem by a sequence of stable problems. Nevertheless, the convergence $\varepsilon_s \to 0$ does not guarantee satisfactory smallness of ε_s for particular ε_d in practice. Sometimes the solutions of linear or nonlinear systems of equations related to inverse problems are very sensitive even in the case of stability. Such examples are presented in Sect. 5.4.2 below.

2.2 Inverse Problems and Non-destructive Evaluation from a Practical Viewpoint

2.2.1 General Remarks

Non-destructive evaluation (NDE) in general terms means analysis and technology for the quantitative characterisation of materials and structures by noninvasive methods, i.e., examination of an object or material without impairing its future usefulness. For this purpose ultrasonic, optic, electromagnetic, and thermographic methods are used. Here we focus on ultrasonic methods only, i.e., examination is carried out with high-frequency mechanical waves.

The use of ultrasound in NDE has been known since the discovery of the piezoelectric effect in quartz in 1880 that enabled electric signals to be used to generate mechanical vibrations. Nowadays ultrasonic methods are widely used for flaw detection, determination of initial or residual stresses in materials, materials characterisation, etc. Ultrasonic diagnosis for medical purposes must also be stressed. There are many of monographs and papers in the field. The earlier results are reflected, for example, in monographs by McGonnagie [49], Truell, Elbaum and Chick [69], Wells [71], Thompson and Chimenti [67] a.o.

An excellent summary on ultrasonic testing of materials and bibliography is given in the monograph by Krautkrämer and Krautkrämer [43]. More recent treatises are by Hauk [31], Shull [64], Kundu [44] and Delsanto [9]. Certainly the list of references above is not exhaustive—only some fundamental treatises are listed.

The practical idea for using ultrasound in NDE is very simple—the waves "feel" the internal structure in the object. By comparing the excited wave (signal) and the measured wavefield after propagating in the object, the needed information will be available. The crucial questions are: (i) what must be measured in order to get this information and (ii) how are the measurements realised in practice.

In principle, the following phenomena can be measured:

(i) flight time of signals, i.e., velocities of various (longitudinal, shear, surface) waves;
(ii) waveform distortion (spectral changes, attenuation, decay, deformation of the surfaces of equal phase, etc.);
(iii) change of polarization of (shear) waves;
(iv) effects of (nonlinear) interaction of waves.

Widely used are the measurements of the flight time of the signals (or wave velocities), which can be used for evaluation of the elastic constants of materials. It means that these measurements link acoustics and elasticity under the name "acoustoelasticity". These studies were intensified around the mid-20th century, stimulated by interest in the evaluation of third order elastic constants (see [5, 61]). Starting from the 1980s of the 20-th century, the interest has also been turned to the nonlinear effects which can considerably enlarge the informative analysis of distorted waveforms—see the overview by Engelbrecht and Ravasoo [13] and the collection of papers edited by Delsanto [9].

2.2.2 Practical Realisation

In what follows, is a very brief description of ultrasonic testing needed for understanding the structure of mathematical models later. The frequency range of the ultrasound used in the NDE is 0.5–15 MHz, occasionally up to 50 MHz. The frequency and the wavelength are related by a simple relation $f \cdot \lambda = c$, where c is the sound velocity. So in steel with $c = 5.9 \cdot 10^3$ m/s, the length for a 10 MHz wave is $\sim$0.6 mm.

In principle, two methods of receiving the ultrasound waves are used [43]: pulse echo method and through-transmission (transit time) method. The schematic setups of both methods are shown in Fig. 2.1.

The next question is related to the main structure of a wavefield generated by an ultrasonic transducer. Again, Krautkrämer and Krautkrämer [43] have analysed

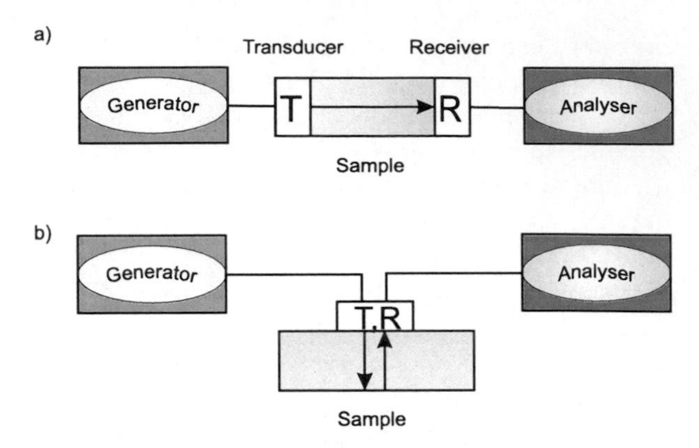

Fig. 2.1 Schematic setups of measurements: (**a**) transit time method, (**b**) pulse-echo method

this question in detail. Classical piezo-electric materials and their properties are described and then the attention is focused on the behaviour of piezo-electric transducers which are widely used in applications. In many cases transducers are made of circular piezo-electric disks. The disks are cut from quartz crystals and their properties depend upon the orientation of cuts. The so-called X-cuts generate longitudinal waves, the Y-cuts shear waves (see, for example, [64]). The transducers generate a wave-field which is often called a wave beam. This wave beam has a special pattern: the near-field (Fresnel) and far-field (Fraunhofer) zones (see [43]). In the near-field zone, the edge waves due to the transducer's boundary, and the plane wave generated from the surface of a transducer are combined in a complicated pattern; in the far-field zone the influence of the edge waves has largely died out. This is schematically shown in Fig. 2.2, where Fig. 2.2a shows the fronts of edge and plane waves and Fig. 2.2b demonstrates the pressure changes on the axis due to the combined influence of edge and plane waves. The geometry of a wave beam is schematically

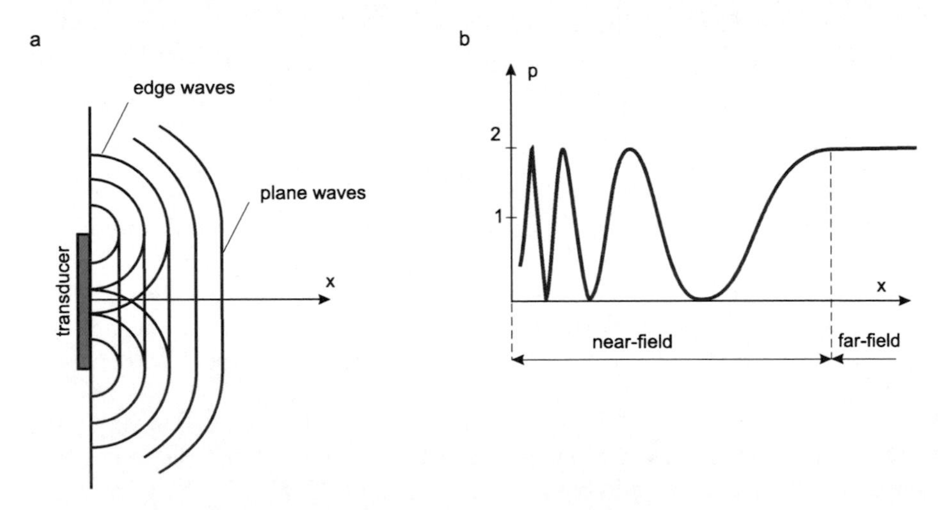

Fig. 2.2 Schematic structured of fields for an ultrasonic transducer: (**a**) near-field with fronts of edge and plane waves; (**b**) the pressure distribution on the axis X

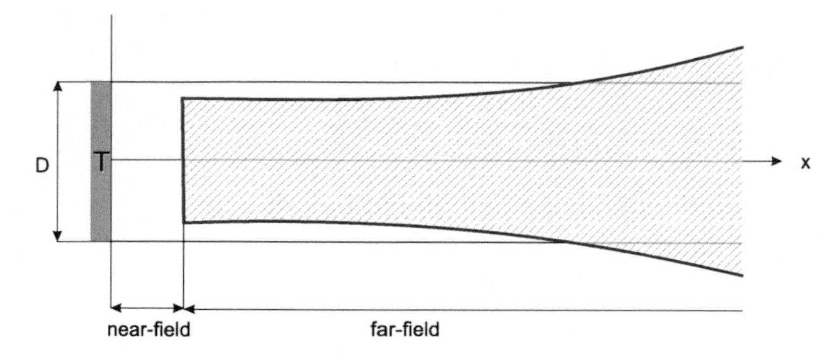

Fig. 2.3 Simplified geometry of fields for an ultrasonic transducer, shaded area shows the essential part of a wave beam

depicted in Fig. 2.3. The essential part of a wave beam can be considered as an 1D wave, influenced by diffractional expansion (beam spreading) in a transverse direction to the beam axis X_1. Its geometry can be changed by using focused transducers [43].

Contemporary technology permits instead of quarts crystals the use of modern piezo-electric materials such as ceramics (barium titanate, lead zirconate titanate, etc.) and polymeric films [64]. In addition, ultrasound may be generated by electromagnetic acoustic transducers or by lasers.

In Sect. 2.1 the inverse problems were briefly analysed from a mathematical viewpoint. Here, in Sect. 2.2 from the viewpoint of realisation, ultrasonic testing is described. In technical terms, the signal generated by a transducer and the signal registered by a receiver are known, and the properties of a sample (a structure) must be determined. It has been demonstrated that a 1D approach might be feasible from the viewpoint of realisation.

Chapter 3
Mathematical Models of Microstructured Solids

3.1 Basic Principles

The conceptual approach in constructing the mathematical models of wave motion is based on the following sequence:

- basic principles agreed (initial assumptions);
- conservation laws formulated;
- constitutive theory constructed (auxiliary postulates introduced in order to form a closed system);
- mathematical models derived (auxiliary assumptions about the character of field variables and approximations of the constitutive equations).

The details of such modelling can be found in monographs by Eringen [23], Eringen and Suhubi [27], Engelbrecht [11], etc. Beside this sequence, certain physical and mathematical requirements are necessary in order to guarantee the best correspondence between the models and reality [26]. These are the following axioms: (i) causality; (ii) determinism; (iii) equipresence; (iv) objectivity; (v) time reversal; (vi) material invariance; (vii) neighbourhood; (viii) memory; (ix) admissibility.

Bearing in mind inverse problems, some of these axioms need explanation. We follow here the formulation given by Eringen and Maugin [26].

Determinism The value of α (a dependent variable) at a material point $\mathbf{X}$ of the body $\mathcal{B}$ at time t is determined by the history of all material points of $\mathcal{B}$.

In terms of inverse problems it means that, in measuring a signal at time t, the information we get may involve not only information at time t, but also historical information. This might be important in the NDT of initial or residual stresses. For NDE of material properties we may restrict ourselves to instantaneous values.

Equipresence At the outset, all constitutive response functionals are to be considered to depend on the same list of constitutive variables, until the contrary is deduced.

J. Janno, J. Engelbrecht, *Microstructured Materials: Inverse Problems*,
Springer Monographs in Mathematics,
DOI 10.1007/978-3-642-21584-1_3,

As said in [26], this is more a precautionary measure rather than an axiom. Still, in proper modelling the equipresence is of utmost importance, because it may reveal interactions between various waves.

Admissibility Constitutive equations must be consistent with the balance laws and the entropy inequality.

Even this short overview demonstrates clearly that the proper modelling is needed in order to get good results in NDE. The mathematical models should be derived on the basis of continuum mechanics contrary to simplified approaches. We do not support the idea to "improve" models by adding a term or two to governing equations as it is sometimes done in simplified applications.

3.2 Microstructured Solids

The classical theory of continuous media is built up using the assumption of smoothness of continua. So the linear theory of elasticity has as its basis the following equations of motion in terms of a displacement U_K, $K = 1, 2, 3$:

$$\rho_0 U_{I,tt} - (\lambda + \mu) U_{K,KI} - \mu U_{I,KK} = 0, \tag{3.1}$$

where ρ_0 is the density and λ, μ are the Lamé constants (the second order elastic moduli). The comma denotes, as usual, differentiation with respect to time t or with respect to the Lagrangian coordinate X_K. The summation rule over repeated indices K and I is used. This model involves material parameters ρ_0, λ and μ, and the velocities of longitudinal and transverse waves c_0 and c_t are calculated by $c_0^2 = (\lambda + 2\mu)/\rho_0, c_t^2 = \mu/\rho_0$. This means that the combination $\lambda + 2\mu$ of Lamé parameters has a crucial importance.

There are two aspects that bring us to more advanced theories of continua than the theory of the homogeneous materials.

First, materials used in contemporary advanced technologies are often characterised by their complex structure satisfying many requirements in practice. This concerns polycrystalline solids, ceramic composites, alloys, functionally graded materials, granular materials, etc. Often the damage effects should also be accounted for, i.e., materials are still usable when they have microcracks. In all these materials there exists an intrinsic space-scale, like the lattice spacing, the size of a crystallite or a grain, or the distance between the microcracks. Clearly the dynamical behaviour of such microstructured materials cannot be explained by the classical theory of homogeneous continua.

Second, all the materials have an internal structure down to molecules or atoms (see, for example, [28]). If loading is of high-frequency, then the wavelengths become very small and the waves start to "feel" the internal structure and again the assumption about the homogeneity is violated.

Within the theories of continua the problems of irregularities of media were actually predicted already by the Cosserats and Voigt, and more recently by Mindlin [50], Eringen [24] and others. The elegant mathematical theories of continua with voids or with vector microstructure, of continua with spins, of micromorphic continua, ferroelectric crystals etc. have been elaborated since; see the overviews by Capriz [4] and Eringen [25]. An excellent overview on the complexity of wave motion was presented by Pastrone [53], see also [19].

The straightforward modelling of microstructured solids leads to assigning all the physical properties to every volume element dV in a solid, thus introducing the dependence on material coordinates X_K. Then the governing equations implicitly include space-dependent parameters but, due to the complexity of the system, can be solved only numerically. Another probably much more effective way is to separate macro- and microstructure in continua. Then the conservation laws for both structures should be separately formulated [24, 25, 50], or the microstructural quantities are separately taken into account in one set of conservation laws [48]. In the first case macrostress and microstress together with the interactive force between macro- and microstructure need to be determined. The last case uses the concept of pseudomomentum and material inhomogeneity force.

Here we follow Mindlin [50] who has interpreted the microstructure "as a molecule of a polymer, a crystallite of a polycrystal or a grain of a granular material". This microelement is taken as a deformable cell. Note that if this cell is rigid, then the Cosserat model follows. The displacement $\mathbf{U}$ of a material particle in terms of macrostructure is defined by its components $U_I = x^I - X_I$, where x^I, X_I $(I = 1, 2, 3)$ are the components of the spatial and material position vectors, respectively. Within each material volume there is a microelement, and the microdisplacement $\mathbf{U}'$ is defined by its components $U'_I \equiv \widehat{x}_I - \widehat{X}_I$, where the origin of the coordinates $\widehat{X}_I$ moves with the displacement $\mathbf{U}$. The displacement gradient is assumed to be small. This leads to the basic assumption of Mindlin [50] that "the microdisplacement can be expressed as a sum of products of specified functions of $\hat{X}_I$ and arbitrary functions of x_I and t". The first approximation is then

$$U'_J = \widehat{x}_K \varphi_{KJ}(x_{I,t}). \tag{3.2}$$

The microdeformation is then

$$\partial U'_J / \partial \widehat{x}_I = \partial'_I U'_J = \varphi_{IJ}. \tag{3.3}$$

Further we consider the simplest 1D case and drop the indices I, J to deal with U and φ only. The indices X, t used in the sequel denote differentiation.

The fundamental balance laws for microstructured material can be formulated separately for macroscopic and microscopic scales. We show here how the balance laws can be derived from the Lagrangian [50, 53] $\mathcal{L} = \mathrm{K} - \mathrm{W}$ formed from the kinetic and potential energies

$$\mathrm{K} = \frac{1}{2}\rho_0 U_t^2 + \frac{1}{2} I \varphi_t^2, \qquad \mathrm{W} = \mathrm{W}(U_X, \varphi, \varphi_X), \tag{3.4}$$

where I is the microinertia related to a microelement.

The corresponding Euler–Lagrange equations have the general form

$$\left(\frac{\partial \mathcal{L}}{\partial U_t}\right)_t + \left(\frac{\partial \mathcal{L}}{\partial U_X}\right)_X - \frac{\partial \mathcal{L}}{\partial U} = 0, \tag{3.5}$$

$$\left(\frac{\partial \mathcal{L}}{\partial \varphi_t}\right)_t + \left(\frac{\partial \mathcal{L}}{\partial \varphi_X}\right)_X - \frac{\partial \mathcal{L}}{\partial \varphi} = 0. \tag{3.6}$$

Inserting the partial derivatives

$$\frac{\partial \mathcal{L}}{\partial U_t} = \rho_0 U_t, \qquad \frac{\partial \mathcal{L}}{\partial U_X} = -\frac{\partial \mathrm{W}}{\partial U_X}, \qquad \frac{\partial \mathcal{L}}{\partial U} = 0, \tag{3.7}$$

$$\frac{\partial \mathcal{L}}{\partial \varphi_t} = I\varphi_t, \qquad \frac{\partial \mathcal{L}}{\partial \varphi_X} = -\frac{\partial \mathrm{W}}{\partial \varphi_X}, \qquad \frac{\partial \mathcal{L}}{\partial \varphi} = -\frac{\partial W}{\partial \varphi}, \tag{3.8}$$

into (3.5), (3.6) we obtain the equations of motion

$$\rho_0 U_{tt} - \left(\frac{\partial \mathrm{W}}{\partial U_X}\right)_X = 0, \qquad I\varphi_{tt} - \left(\frac{\partial \mathrm{W}}{\partial \varphi_X}\right)_X + \frac{\partial \mathrm{W}}{\partial \varphi} = 0. \tag{3.9}$$

Denoting

$$T = \frac{\partial \mathrm{W}}{\partial U_X}, \qquad P = \frac{\partial \mathrm{W}}{\partial \varphi_X}, \qquad R = \frac{\partial \mathrm{W}}{\partial \varphi}, \tag{3.10}$$

we recognise $T = T^{11}$ as the macrostress (the first Piola–Kirchhoff stress), P as the microstress and R as the interactive force. The equations of motion (3.9) now take the form

$$\rho_0 U_{tt} = T_X, \qquad I\varphi_{tt} = P_X - R. \tag{3.11}$$

The simplest potential energy function describing the influence of a microstructure is a quadratic function

$$\mathrm{W} = \frac{1}{2}aU_X^2 + A\varphi U_X + \frac{1}{2}B\varphi^2 + \frac{1}{2}C\varphi_X^2, \tag{3.12}$$

with a, A, B, C denoting material constants. Inserting it into (3.10) and the result into (3.11), the governing equations take the form

$$\rho_0 U_{tt} = aU_{XX} + A\varphi_X, \tag{3.13}$$

$$I\varphi_{tt} = C\varphi_{XX} - AU_X - B\varphi. \tag{3.14}$$

This is the sought mathematical model for 1D longitudinal waves in microstructured materials of the Mindlin type. Due to the physical background, the coefficients of the system (3.13), (3.14) satisfy the inequalities

$$\rho_0, a, I, C, B > 0. \tag{3.15}$$

Now we introduce physical nonlinearity. Instead of the potential energy function (3.12) we use $W = W_2 + W_3$ where W_2 is the simplest quadratic function

$$W_2 = \frac{1}{2}aU_X^2 + A\varphi U_X + \frac{1}{2}B\varphi^2 + \frac{1}{2}C\varphi_X^2, \tag{3.16}$$

and W_3 includes cubic terms, i.e., nonlinearities on both the macro- and microlevel:

$$W_3 = \frac{1}{6}NU_X^3 + \frac{1}{6}M\varphi_X^3, \tag{3.17}$$

where N and M are constants (see, for example, [5]). We neglect here geometrical nonlinearity because its effect for the conventional materials is small [11, 12]. Using again (3.10) and (3.11), we obtain

$$\rho_0 U_{tt} = aU_{XX} + NU_X U_{XX} + A\varphi_X, \tag{3.18}$$

$$I\varphi_{tt} = C\varphi_{XX} + M\varphi_X\varphi_{XX} - AU_X - B\varphi. \tag{3.19}$$

The derived linear (3.13), (3.14) and nonlinear (3.18), (3.19) systems reflect the coupling of macro- and microdeformations. This coupling is characterised by a certain hierarchical sequence which can be better revealed by asymptotic analysis. For this we shall use dimensionless variables.

Let us introduce first

$$x = \frac{X}{L}, \qquad \hat{t} = \frac{t}{T_0}, \qquad u = \frac{U}{U_0}, \tag{3.20}$$

where U_0, L and T_0 are certain constants (e.g. amplitude, wavelength and period of an initial excitation). Then we introduce geometric parameters

$$\delta = \frac{l^2}{L^2}, \qquad \varepsilon = \frac{U_0}{L}, \qquad \varkappa = \frac{T_0^2}{L^2}, \tag{3.21}$$

where l is the scale of the microstructure. We start from the nonlinear system (3.18), (3.19) and later present the linear equivalent from the derivation. Substituting (3.20) and (3.21) into system (3.18), (3.19), we obtain

$$\rho_0 u_{tt} = a\varkappa u_{xx} + N\varkappa\varepsilon u_x u_{xx} + \frac{A\varkappa}{\varepsilon}\varphi_x, \tag{3.22}$$

$$\delta I^*\varphi_{tt} = \delta C^*\varphi_{xx} + \delta^{3/2}M^*\varphi_x\varphi_{xx} - A\varepsilon u_x - B\varphi. \tag{3.23}$$

Here we neglected the tilde over t (i.e. time is dimensionless) and introduced scaling constants by

$$I^* = \frac{I}{\varkappa l^2}, \qquad C^* = \frac{C}{l^2}, \qquad M^* = \frac{M}{l^3}. \tag{3.24}$$

Now we are going to deduce a simplified approximate model. To this end we eliminate microdeformation φ from (3.22), (3.23) making use of the *slaving principle* (cf. [14, 34, 54]). We deduce from (3.23) the expression for φ

$$\varphi = -\frac{A\varepsilon}{B}u_x + \frac{\delta}{B}\left(C^*\varphi_{xx} - I^*\varphi_{tt}\right) + \frac{\delta^{3/2}M^*}{B}\varphi_x\varphi_{xx}, \tag{3.25}$$

and expand φ into a Taylor series with respect to $\delta^{1/2}$

$$\varphi = \varphi_0 + \delta^{1/2}\varphi_1 + \delta\varphi_2 + \delta^{3/2}\varphi_3 + \cdots. \tag{3.26}$$

Then we obtain the following formulae for the first four terms in this expansion:

$$\varphi_0 = -\frac{A\varepsilon}{B}u_x, \qquad \varphi_1 = 0, \tag{3.27}$$

$$\varphi_2 = \frac{A\varepsilon}{B^2}\left(I^*u_{tt} - C^*u_{xx}\right)_x, \tag{3.28}$$

$$\varphi_3 = \frac{A^2M^*\varepsilon^2}{2B^3}\left(u_{xx}^2\right)_x. \tag{3.29}$$

Substituting

$$\varphi_0 + \delta\varphi_2 + \delta^{3/2}\varphi_3 \tag{3.30}$$

for φ into (3.22) we arrive at the following hierarchical governing equation for u:

$$u_{tt} = bu_{xx} + \frac{\mu}{2}\left(u_x^2\right)_x + \delta(\beta u_{tt} - \gamma u_{xx})_{xx} + \delta^{3/2}\frac{\lambda}{2}\left(u_{xx}^2\right)_{xx}, \tag{3.31}$$

where

$$b = \frac{a\varkappa}{\rho_0}\left(1 - \frac{A^2}{aB}\right), \qquad \mu = \frac{N\varkappa\varepsilon}{\rho_0},$$
$$\beta = \frac{A^2\varkappa I^*}{B^2\rho_0}, \qquad \gamma = \frac{A^2\varkappa C^*}{B^2\rho_0}, \qquad \lambda = \frac{A^3M^*\varkappa\varepsilon}{B^3\rho_0}. \tag{3.32}$$

Equation (3.31) involves two wave operators

$$u_{tt} - bu_{xx} - \frac{\mu}{2}\left(u_x^2\right)_x, \tag{3.33}$$

$$\delta\left(\beta u_{tt} - \gamma u_{xx} + \delta^{1/2}\frac{\lambda}{2}u_{xx}^2\right)_{xx}, \tag{3.34}$$

characteristic of macro- and microstructure, respectively. If the scale parameter δ is small then the influence of microstructure can be neglected. Conversely, if δ is large then the influence of macrostructure is weaker and the wave process is governed by the properties of microstructure. Clearly, the intermediate case includes both effects. The similar procedure for linear system (3.13), (3.14) yields

$$u_{tt} = bu_{xx} + \delta(\beta u_{tt} - \gamma u_{xx})_{xx}, \tag{3.35}$$

which certainly reveals the same hierarchical features with linear wave operators.

In terms of the deformation $v = u_x$, (3.31) reads

$$v_{tt} = bv_{xx} + \frac{\mu}{2}\left(v^2\right)_{xx} + \delta(\beta v_{tt} - \gamma v_{xx})_{xx} + \delta^{3/2}\frac{\lambda}{2}\left(v_x^2\right)_{xxx}. \tag{3.36}$$

We note that the inequalities

$$0 < b < \frac{a\varkappa}{\rho_0}, \qquad \delta, \beta, \gamma > 0, \tag{3.37}$$

are valid for the coefficients b, δ, β and γ. Indeed, the relation $b > 0$ is the necessary hyperbolicity condition and other inequalities in (3.37) follow from the physical conditions (3.15) and the definitions (3.32), (3.24). The equation for v (3.36) can be complemented by the explicit formula for φ deduced from (3.30) by means of (3.27)–(3.29) and (3.32)

$$\varphi = \frac{1}{\vartheta_0}\left[(b - a_0)v + \delta(\beta v_{tt} - \gamma v_{xx}) + \delta^{3/2}\frac{\lambda}{2}\left(v_x^2\right)_x\right], \tag{3.38}$$

where

$$a_0 = \frac{a\varkappa}{\rho_0}, \qquad \vartheta_0 = \frac{A\varkappa}{\varepsilon\rho_0}. \tag{3.39}$$

The mathematical models derived above involve higher order derivatives which model dispersive effects. This is the main advantage of such models. The analysis is based on two balance laws of momentum but it is possible to show that the same result will be obtained by using the concept of pseudomomentum [48]. This has been shown by, for example in [18]. Moreover, the concept of dual internal variables leads also to same result [3]. Based on this concept, it is possible to generalise many known mathematical models from a unified viewpoint.

3.3 General Formulation of Inverse Problems

We are going to study the reconstruction of parameters of microstructured materials from measurements of ultrasonic wave signals (NDE). These parameters appear as coefficients of the differential equations derived in the previous section. We will treat inverse problems for two models:

(A) Problems for the *hierarchical equation* (3.36). Our aim is to determine 5 unknown coefficients b, μ, β, γ and λ of this equation. In the linear case when

$\mu = \lambda = 0$, the number of unknowns reduces to 3, i.e., we have to find only b, β and γ. In all problems the geometrical parameter δ is assumed to be known. Otherwise it is possible to reconstruct the products of the form $\delta\beta$, $\delta\gamma$ and $\delta^{3/2}\lambda$ instead of β, γ and λ.

(B) Problems to determine the coefficients of the coupled system for u and ψ. We follow the nonlinear homogeneous equations (3.22), (3.23) and modify them in order to get a proper system for the inverse problem. The reason for such a modification is that the coefficients of (3.22), (3.23) are not uniquely recovered by the solution pair (u, φ). Indeed, any vector of coefficients that fits to this system can be multiplied by an arbitrary constant to get another vector of coefficients that also fits to this system. The reconstruction of all coefficients could be possible only in the case of a non-homogeneous system involving mass forces.

Therefore, we divide (3.22) by ρ_0 and (3.23) by I^* to get coefficients 1 at time derivatives. In deformation variables the resulting system reads

$$v_{tt} = a_0 v_{xx} + \frac{\mu}{2}\left(v^2\right)_{xx} + \vartheta_0 \varphi_{xx}, \tag{3.40}$$

$$\delta\varphi_{tt} = \delta a_1 \varphi_{xx} + \delta^{3/2} \nu_1 \varphi_x \varphi_{xx} - \alpha\varphi - \vartheta_1 v \tag{3.41}$$

where $a_0 = \frac{a\varkappa}{\rho_0}$, $\mu = \frac{N\varkappa\varepsilon}{\rho_0}$, $\vartheta_0 = \frac{A\varkappa}{\varepsilon\rho_0}$, as before, and

$$a_1 = \frac{C^*}{I^*}, \qquad \nu_1 = \frac{M^*}{I^*}, \qquad \alpha = \frac{B}{I^*}, \qquad \vartheta_1 = \frac{A\varepsilon}{I^*}. \tag{3.42}$$

Throughout the book we call (3.40), (3.41) the *coupled system*. We are going to study inverse problems to determine 7 unknown coefficients a_0, μ, ϑ_0, a_1, ν_1, α and ϑ_1 of this system. In the linear case when $\mu = \nu_1 = 0$, the number of unknowns reduces to 5.

To determine the unknown coefficients, we make use of harmonic waves and wave packets in linear cases and solitary waves in the nonlinear cases. The inverse problems for the approximate hierarchical equation are easier to treat that the corresponding problems for the coupled system. We study both models in parallel, starting with the easier one and continuing with the more complex one.

In the coupled system two state variables v and φ are involved. Nevertheless, in practice it is realistic to measure only the marcodeformation v. As we will see later on, this brings along additional identifiability restrictions. Namely, the linear waves in macro-level do not contain enough information to reconstruct both ϑ_0 and ϑ_1. In this case we are able to determine 4 quantities: a_0, a_1, α and

$$\vartheta = \vartheta_0 \vartheta_1.$$

In the nonlinear case we can determine 6 coefficients: a_0, a_1, α, ϑ, μ and

$$\nu = \frac{\nu_1}{\vartheta_0}$$

from solitary waves in macro-level. However, when the registration of micro-waves is possible, the parameters ϑ_0, ϑ_1 and ν_1 can also be extracted.

The coefficients of the coupled system satisfy the following a priori inequalities

$$a_0, a_1, \alpha, \vartheta > 0. \qquad (3.43)$$

They easily follow from (3.39) and (3.42) in view of the physical inequalities (3.15) and the definitions (3.24). Further, in the case when the scale of the microstructure is zero, i.e., $\delta = 0$, from (3.41) we get $\varphi = -\frac{\vartheta_1}{\alpha} v$. Plugging this relation into (3.40) we reach the equation $v_{tt} = (a_0 - \frac{\vartheta}{\alpha}) v_{xx} + \frac{\mu}{2}(v^2)_{xx}$ for the macrodeformation. From this equation we infer the following necessary hyperbolicity condition for the coefficients:

$$a_0 \alpha - \vartheta > 0. \qquad (3.44)$$

The solutions of inverse problems in two models under consideration are related to each other as follows (cf. (3.32), (3.39) and (3.42)):

$$b = a_0 - \frac{\vartheta}{\alpha}, \qquad \beta = \frac{\vartheta}{\alpha^2}, \qquad \gamma = \frac{\vartheta a_1}{\alpha^2}, \qquad \lambda = \frac{\vartheta^2 \nu}{\alpha^3}. \qquad (3.45)$$

The parameter μ is the same in both models.

Chapter 4
Linear Waves

4.1 Dispersion Relations. Harmonic Waves

In this chapter we investigate the solutions of the hierarchical equation and the coupled system in the linear case. The simplest linear wave is harmonic. *Harmonic waves* in the models under consideration have the form

$$v(x,t) = Ae^{i(kx-\omega t)}, \qquad \varphi(x,t) = \widehat{A}e^{i(\widehat{k}x-\widehat{\omega}t)} \tag{4.1}$$

where ω, k, A and $\widehat{\omega}$, $\widehat{k}$, $\widehat{A}$ are the frequencies, the wavenumbers and the amplitudes of the macro- and microwaves, respectively. As we will see later on, both our models (i.e. the hierarchical equation and the coupled system) may have only synchronous harmonic waves. This means that $\widehat{\omega} = \omega$ and $\widehat{k} = k$.

Plugging the solution formula (4.1) into a governing equation (or system) leads to an algebraic relation for ω and k. This is called the *dispersion equation*. In the following subsections we will discuss this issue in detail.

4.1.1 Hierarchical Equation

The hierarchical equation (3.36) in the linear case has the form

$$v_{tt} = bv_{xx} + \delta(\beta v_{tt} - \gamma v_{xx})_{xx} \tag{4.2}$$

and the related formula for φ (3.38) with the neglected nonlinear term is

$$\varphi = \frac{1}{\vartheta_0}\big[(b-a_0)v + \delta(\beta v_{tt} - \gamma v_{xx})\big]. \tag{4.3}$$

Using in (4.3) $v = Ae^{i(kx-\omega t)}$, we get the formula for synchronous φ

$$\varphi = \widehat{A}e^{i(kx-\omega t)} \quad \text{with } \widehat{A} = \frac{A}{\vartheta_0}\big[b - a_0 + \delta\big(k^2 - \omega^2\big)\big]. \tag{4.4}$$

J. Janno, J. Engelbrecht, *Microstructured Materials: Inverse Problems*,
Springer Monographs in Mathematics,
DOI 10.1007/978-3-642-21584-1_4,

On the other hand, the substitution of v by $Ae^{i(kx-\omega t)}$ in the partial differential equation (4.2) results in the following dispersion equation:

$$\delta\beta\omega^2k^2 - \delta\gamma k^4 + \omega^2 - bk^2 = 0. \tag{4.5}$$

We can solve the algebraic equation (4.5) both for ω and k. The solution with respect to the frequency is $\omega = \pm\omega(k)$ where

$$\omega(k) = k\sqrt{\frac{b+\delta\gamma k^2}{1+\delta\beta k^2}}. \tag{4.6}$$

Owing to the basic inequalities (3.37), the function $\omega(k)$ is real for any real k. The related harmonic waves $Ae^{i(kx-\omega(k)t)}$ and $Ae^{i(kx+\omega(k)t)}$ propagate to the right and left, respectively, with the phase velocities

$$c_{ph} = \frac{\omega}{k} = \pm\sqrt{\frac{b+\delta\gamma k^2}{1+\delta\beta k^2}}. \tag{4.7}$$

From the inverse problems' viewpoint the solution of the dispersion equation for the wavenumber is of more interest because this appears in frequency decompositions of time-series of data of the problems. Therefore, let us solve (4.5) for k, too. We obtain four solution branches $k = \pm k(\omega)$ and $k = \pm k_2(\omega)$ where

$$k(\omega) = \omega\sqrt{\frac{1}{2\delta\gamma\omega^2}\left[\delta\beta\omega^2 - b + \sqrt{\left(\delta\beta\omega^2 - b\right)^2 + 4\delta\gamma\omega^2}\right]}, \tag{4.8}$$

$$k_2(\omega) = \omega\sqrt{\frac{1}{2\delta\gamma\omega^2}\left[\delta\beta\omega^2 - b - \sqrt{\left(\delta\beta\omega^2 - b\right)^2 + 4\delta\gamma\omega^2}\right]}. \tag{4.9}$$

The solutions $k(\omega)$ and $k_2(\omega)$ represent the *acoustic* and *optical branch* of the dispersion function, respectively. In view of the inequalities (3.37) we immediately see that for any real ω the function $k(\omega)$ is real and $k_2(\omega)$ is imaginary. Therefore, only the acoustic branch yields harmonic waves $Ae^{i(k(\omega)x-\omega t)}$ and $Ae^{i(-k(\omega)x-\omega t)}$.

The inverse of $k(\omega)$ is the function $\omega(k)$ given by (4.6). By means of elementary mathematical analysis the following basic properties of $k(\omega)$ can be obtained: $k(\omega)$ is strictly increasing and

$$k(\omega) \sim \frac{1}{\sqrt{b}}\omega \quad \text{as } \omega \to 0, \qquad k(\omega) \sim \sqrt{\frac{\beta}{\gamma}}\omega \quad \text{as } \omega \to \pm\infty. \tag{4.10}$$

Further, let us establish the dependence of the type of the dispersion on the coefficients. For this purpose, we have to compare the phase velocity $c_{ph} = \frac{\omega}{k(\omega)}$ with the group velocity $c_g = \omega'(k) = \frac{1}{k'(\omega)}$. The following lemma holds, the proof being given in Sect. 4.3.

Lemma 4.1 (1) *In case* $b\beta - \gamma > 0$ *the inequality* $c_g < c_{ph}$ *is valid for any* $\omega \in \mathbb{R}$, *which means that the model possesses the normal dispersion.*

(2) *In case* $b\beta - \gamma < 0$ *the inequality* $c_g > c_{ph}$ *is valid for any* $\omega \in \mathbb{R}$, *which means that the anomalous dispersion occurs.*

(3) *In case* $b\beta - \gamma = 0$ *the equality* $c_g = c_{ph}$ *holds for any* $\omega \in \mathbb{R}$. *Then, the material is nondispersive and the function* $k(\omega)$ *is linear:* $k(\omega) = \frac{1}{\sqrt{b}}\omega$.

4.1.2 Coupled System

Now we study the coupled system (3.40), (3.41) in the linear case, i.e., the system

$$v_{tt} = a_0 v_{xx} + \vartheta_0 \varphi_{xx}, \tag{4.11}$$

$$\delta\varphi_{tt} = \delta a_1 \varphi_{xx} - \alpha\varphi - \vartheta_1 v. \tag{4.12}$$

We seek the solutions (v, φ) of this system whose first component v has the form of the harmonic wave $v(x,t) = Ae^{i(kx-\omega t)}$. Plugging this formula of v into (4.11) we immediately get the following equation of φ:

$$\vartheta_0 \varphi_{xx} = -A\big(\omega^2 - a_0 k^2\big)e^{i(kx-\omega t)}.$$

The solution of this equation among the harmonic waves is

$$\varphi = \widehat{A}e^{i(kx-\omega t)} \quad \text{with } \widehat{A} = \frac{A(\omega^2 - a_0 k^2)}{\vartheta_0 k^2}. \tag{4.13}$$

This implies that v and φ are synchronous.

Further, plugging $v = Ae^{i(kx-\omega t)}$ and $\varphi = \widehat{A}e^{i(kx-\omega t)}$ into (4.12), using the formula of $\widehat{A}$, dividing by $Ae^{i(kx-\omega t)}$ and simplifying, we obtain the following quartic dispersion equation:

$$\omega^4 + \varkappa_1\omega^2 k^2 + \varkappa_2 k^4 + \varkappa_3\omega^2 + \varkappa_4 k^2 = 0 \tag{4.14}$$

where

$$\varkappa_1 = -(a_0 + a_1), \qquad \varkappa_2 = a_0 a_1, \qquad \varkappa_3 = -\frac{\alpha}{\delta}, \qquad \varkappa_4 = \frac{a_0\alpha - \vartheta}{\delta} \tag{4.15}$$

and $\vartheta = \vartheta_0\vartheta_1$, as defined in Sect. 3.3.

For a given frequency ω, equation (4.14) has four solutions $k = \pm k(\omega)$ and $k = \pm k_2(\omega)$, where

$$k(\omega) = \omega\sqrt{\frac{a_0 + a_1 - \frac{a_0\alpha - \vartheta}{\delta\omega^2} + \sqrt{(a_0 - a_1 - \frac{a_0\alpha-\vartheta}{\delta\omega^2})^2 + \frac{4a_1\vartheta}{\delta\omega^2}}}{2a_0 a_1}}, \tag{4.16}$$

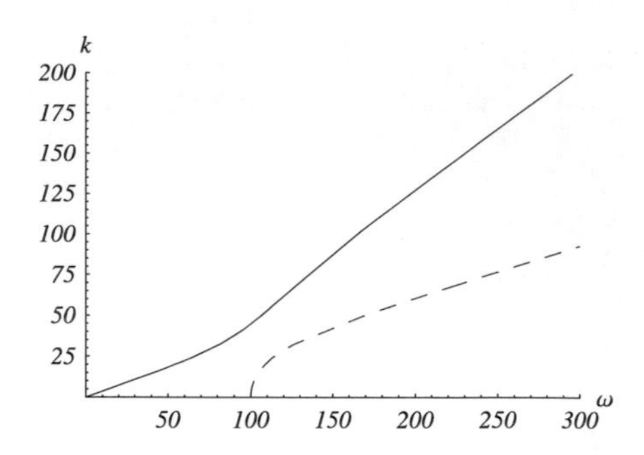

Fig. 4.1 Functions $k(\omega)$ (*solid line*) and $k_2(\omega)$ (*dashed line*) in case $a_0 = 10$, $a_1 = 2$, $\alpha = 1$, $\vartheta = 2$ and $\delta = 10^{-4}$

$$k_2(\omega) = \omega\sqrt{\frac{a_0 + a_1 - \frac{a_0\alpha-\vartheta}{\delta\omega^2} - \sqrt{(a_0 - a_1 - \frac{a_0\alpha-\vartheta}{\delta\omega^2})^2 + \frac{4a_1\vartheta}{\delta\omega^2}}}{2a_0a_1}}. \tag{4.17}$$

To analyse these branches, we use the physical inequalities (3.43) and (3.44). Then we easily see that the function $k(\omega)$ is real for any real ω and satisfies $k(0) = 0$. This means that $k(\omega)$ represents the acoustic branch. But unlike the case of the hierarchical equation, the optical branch $k_2(\omega)$ now has real values, too. More precisely, $k_2(\omega)$ is imaginary for $\omega \in \mathbb{R}$ such that $|\omega| < \sqrt{\frac{\alpha}{\delta}}$, real for $\omega \in \mathbb{R}$ such that $|\omega| \geq \sqrt{\frac{\alpha}{\delta}}$ and $k_2(\pm\sqrt{\frac{\alpha}{\delta}}) = 0$. We emphasise that real values of $k_2(\omega)$ occur only at bigger values of the frequency, because δ is a small number. The acoustic and optical branches are illustrated in Fig. 4.1.

Summing up, given a frequency ω four types of harmonic waves may occur: $Ae^{i(k(\omega)x-\omega t)}$, $Ae^{i(-k(\omega)x-\omega t)}$, and in case $|\omega| \geq \sqrt{\frac{\alpha}{\delta}}$ also $Ae^{i(k_2(\omega)x-\omega t)}$, $Ae^{i(-k_2(\omega)x-\omega t)}$.

By means of the elementary techniques of the mathematical analysis it can be verified that $k(\omega)$ and $k_2(\omega)$ are strictly increasing and

$$\begin{aligned}
k(\omega) &\sim \sqrt{\frac{\alpha}{a_0\alpha - \vartheta}}\,\omega \quad \text{as } \omega \to 0, \\
k(\omega) &\sim \max\left\{\frac{1}{\sqrt{a_0}}; \frac{1}{\sqrt{a_1}}\right\}\omega \quad \text{as } |\omega| \to \infty, \\
k_2(\omega) &\sim \min\left\{\frac{1}{\sqrt{a_0}}; \frac{1}{\sqrt{a_1}}\right\}\omega \quad \text{as } |\omega| \to \infty.
\end{aligned} \tag{4.18}$$

Finally, we deal with the type of the dispersion. In the present case we have to consider two cases: acoustic waves with the phase and group velocities $c_{ph} = \frac{\omega}{k(\omega)}$, $c_g = \frac{1}{k'(\omega)}$ and optical waves with the phase and group velocities $c_{ph2} = \frac{\omega}{k_2(\omega)}$, $c_{g2} = \frac{1}{k_2'(\omega)}$. The following lemma (the proof is in Sect. 4.3 again) gives the dependence of the type of the dispersion on the coefficients.

Lemma 4.2 (1) *In case* $a_0\alpha - a_1\alpha - \vartheta > 0$ *the inequality* $c_{ph} > c_g$ *is valid for any* $\omega \in \mathbb{R}$. *This means that acoustic waves have the normal dispersion.*

(2) *In case* $a_0\alpha - a_1\alpha - \vartheta < 0$ *the inequality* $c_{ph} < c_g$ *is valid for any* $\omega \in \mathbb{R}$. *Hence acoustic waves have the anomalous dispersion.*

(3) *In case* $a_0\alpha - a_1\alpha - \vartheta = 0$ *the equality* $c_{ph} = c_g$ *holds for any* $\omega \in \mathbb{R}$. *Acoustic waves are nondispersive. Then the function k is linear:* $k(w) = \frac{1}{\sqrt{a_1}} w$ *and* k_2 *has the form* $k_2(\omega) = \omega\sqrt{\frac{1}{a_0}(1 - \frac{\alpha}{\delta\omega^2})}$.

(4) *Optical waves always have the normal dispersion, i.e.,* $c_{ph2} > c_{g2}$ *for any* $a_0, a_1, \alpha, \vartheta$ *and* $|\omega| \geq \sqrt{\frac{\alpha}{\delta}}$.

Convention In the sequel the phrases dispersive case, nondispersive case, normal dispersion and anomalous dispersion always mean the types of dispersion that are related to acoustic waves.

The case of the anomalous dispersion contains two important subcases: *weak anomaly* $a_1 < a_0 < a_0 + \frac{\vartheta}{\alpha}$ and *strong anomaly* $a_0 < a_1$. The equality $a_0 = a_1$ occurs between these subcases. We call the latter one *midpoint of the anomaly*.

Some features of direct problems are different in the cases of weak and strong anomaly, e.g. the coincidence of dispersion functions of two models to be discussed in the next subsection, and the ranges of the velocity c of solitary waves presented at the end of Sect. 6.3.3. Two solutions of the inverse problem for harmonic waves coincide at the midpoint $a_0 = a_1$ (Theorem 5.3).

4.1.3 Comparison of Models

In this subsection we compare the dispersion relations in our two models: the hierarchical equation and the coupled system. To this end we make use of the relations (3.45) between the parameters of these models.

The formulas (3.45) imply that $b\beta - \gamma = \frac{\vartheta}{\alpha^3}(a_0\alpha - a_1\alpha - \vartheta)$. Thus, since $\alpha, \vartheta > 0$, we have

$$\operatorname{sign}(b\beta - \gamma) = \operatorname{sign}(a_0\alpha - a_1\alpha - \vartheta).$$

Therefore, the conditions that distinguish the types of the dispersion (normal versus anomalous) in the models under consideration given at the ends of Sects. 4.1.1 and 4.1.2, are exactly the same.

In the nondispersive case the related linear functions $k(\omega) = \frac{1}{\sqrt{b}}\omega$ and $k(\omega) = \frac{1}{\sqrt{a_1}} w$ coincide. However, the nondispersive materials are rather theoretical because the probability of the occurrence of the equality $a_0\alpha - a_1\alpha - \vartheta = 0$ is zero.

Further, let us compare the functions $k(\omega)$ in the case of the presence of the dispersion. Observing the relations (4.10) and (4.18) we see that these functions asymptotically coincide in the process $\omega \to 0$. But in the process $|\omega| \to \infty$ the coincidence occurs only provided $a_0 \geq a_1$ (i.e. when the material possesses either the

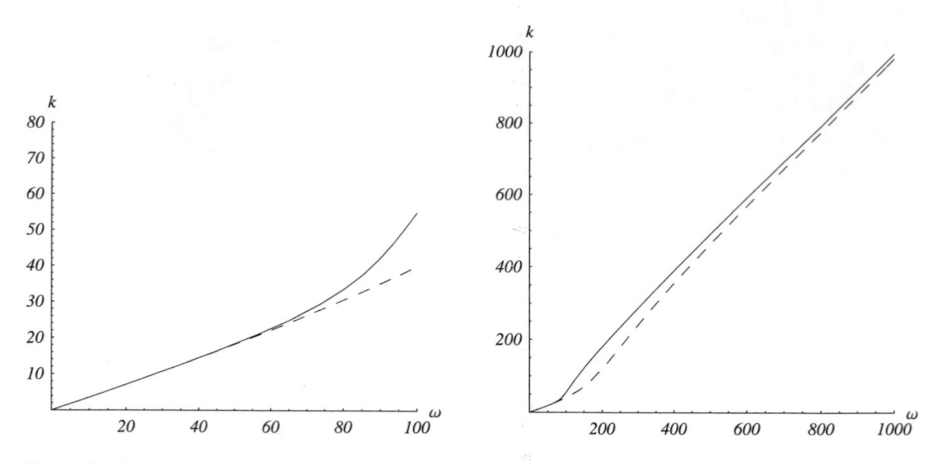

Fig. 4.2 Functions $k(\omega)$ (CS—*solid line*, HE—*dashed line*) in the case of normal dispersion: $a_0 = 10$, $a_1 = 1$, $\alpha = 1$, $\vartheta = 2$ and $\delta = 10^{-4}$. Then $b = 8$, $\beta = \gamma = 2$

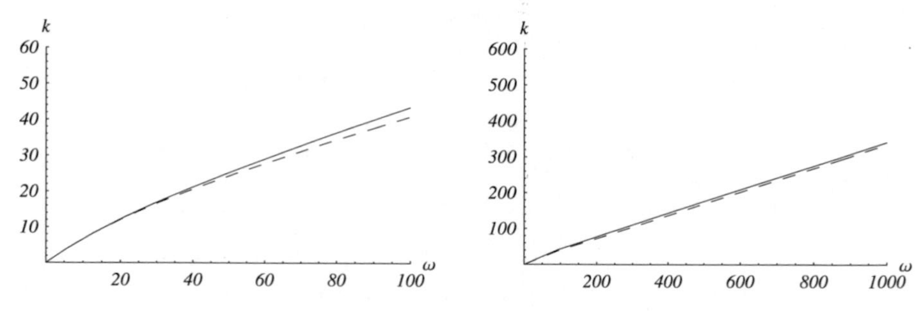

Fig. 4.3 Functions $k(\omega)$ (CS—*solid line*, HE—*dashed line*) in the case of weakly anomalous dispersion: $a_0 = 10$, $a_1 = 9$, $\alpha = 1$, $\vartheta = 8$ and $\delta = 10^{-4}$. Then $b = 2$, $\beta = 8$, $\gamma = 72$

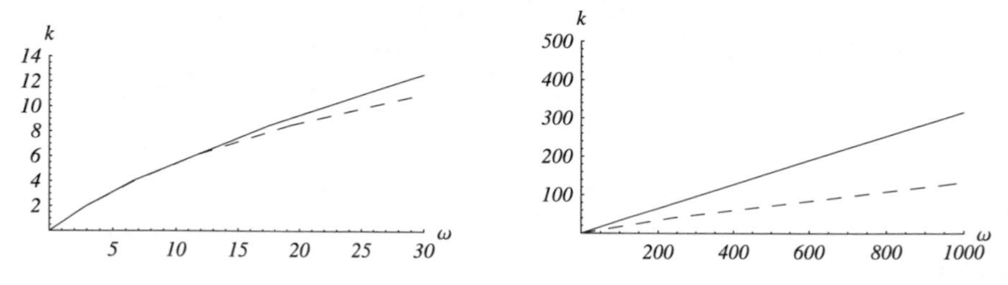

Fig. 4.4 Functions $k(\omega)$ (CS—*solid line*, HE—*dashed line*) in the case of strongly anomalous dispersion: $a_0 = 10$, $a_1 = 60$, $\alpha = 1$, $\vartheta = 8$ and $\delta = 10^{-4}$. Then $b = 2$, $\beta = 8$, $\gamma = 480$

normal or weakly anomalous dispersion). In the opposite case $a_0 < a_1$ (i.e. strongly anomalous dispersion) the functions $k(\omega)$ behave differently at large values of ω.

The behaviour of the functions $k(\omega)$ in the two models is illustrated in Figs. 4.2–4.4. There the abbreviations CS and HE stand for the coupled system and hierarchical equation, respectively. In the cases of normal dispersion and the weak anomalous dispersion the coincidence is good (Figs. 4.2 and 4.3). But in the case of the strong anomaly $a_0 < a_1$ a big difference occurs in large values of ω (Fig. 4.4).

4.2 Other Linear Waves

4.2.1 General Solution Formula

The solution of the linear homogeneous PDEs with constant coefficients (4.2) and (4.11), (4.12) can be performed by standard techniques. For instance, using the Fourier method with respect to the time, we obtain the following general formulas (we complement (4.2) with (4.3)):

$$\begin{aligned} v &= v_+ + v_- + v_{2,+} + v_{2,-}, \\ \varphi &= \varphi_+ + \varphi_- + \varphi_{2,+} + \varphi_{2,-} \end{aligned} \tag{4.19}$$

with

$$\begin{aligned} v_\pm(x,t) &= \frac{1}{2\pi}\int_{-\infty}^{\infty} A_\pm(\omega)e^{i(\pm k(\omega)x-\omega t)}\,d\omega, \\ v_{2,\pm}(x,t) &= \frac{1}{2\pi}\int_{-\infty}^{\infty} A_{2,\pm}(\omega)e^{i(\pm k_2(\omega)x-\omega t)}\,d\omega, \\ \varphi_\pm(x,t) &= \frac{1}{2\pi}\int_{-\infty}^{\infty} \widehat{A}_\pm(\omega)e^{i(\pm k(\omega)x-\omega t)}\,d\omega, \\ \varphi_{2,\pm}(x,t) &= \frac{1}{2\pi}\int_{-\infty}^{\infty} \widehat{A}_{2,\pm}(\omega)e^{i(\pm k_2(\omega)x-\omega t)}\,d\omega. \end{aligned} \tag{4.20}$$

Here $k(\omega)$ and $k_2(\omega)$ are the dispersion functions introduced in the previous section. The addends v_+, φ_+ and v_-, φ_- represent the acoustic waves propagating to the right and left, respectively, and the terms $v_{2,+}$, $\varphi_{2,+}$, $v_{2,-}$, $\varphi_{2,-}$ are the optical components. The coefficients $A_\pm$, $A_{2,\pm}$, $\widehat{A}_\pm$, $\widehat{A}_{2,\pm}$ in the integrands are the spectra of the related components. The macrostructure spectra $A_\pm$, $A_{2,\pm}$ may be arbitrary functions (also singular distributions). But the microstructure spectra $\widehat{A}_\pm$, $\widehat{A}_{2,\pm}$ depend on

$A_\pm$, $A_{2,\pm}$. Namely, they are given by the formulas

$$\begin{aligned} \widehat{A}_\pm(\omega) &= \frac{A_\pm(\omega)}{\vartheta_0}\left[b - a_0 + \delta\left(\left(k(\omega)\right)^2 - \omega^2\right)\right], \\ \widehat{A}_{2,\pm}(\omega) &= \frac{A_{2,\pm}(\omega)}{\vartheta_0}\left[b - a_0 + \delta\left(\left(k_2(\omega)\right)^2 - \omega^2\right)\right] \end{aligned} \tag{4.21}$$

and

$$\widehat{A}_{\pm}(\omega) = \frac{A_{\pm}(\omega)[w^2 - a_0(k(\omega))^2]}{\vartheta_0(k(\omega))^2},$$
$$\widehat{A}^m_{2,\pm}(\omega) = \frac{A_{2,\pm}(\omega)[w^2 - a_0(k_2(\omega))^2]}{\vartheta_0(k_2(\omega))^2} \tag{4.22}$$

in the cases of the hierarchical equation and the coupled system, respectively.

The presented general formulas can be used to extract solutions corresponding to given additional conditions. In the forthcoming subsections we deduce solutions for some important particular problems. In this connection we will deal mainly with the macro-component v.

4.2.2 Right-Propagating Waves

We consider a semi-infinite body that is located to the right of the plane $x = 0$ under the assumption that it is in the condition of equilibrium for negative time values. This means that $v = 0$ for $t \leq 0$. Firstly, we study the problem with the specified deformation at $x = 0$, i.e.,

$$v(0, t) = g(t) \tag{4.23}$$

where g is a known function. The wave caused by such a perturbation propagates to the right. Therefore, the term v_- in (4.19) vanishes. Moreover, in the case of the *hierarchical equation*, the optical branch of the dispersion function $k_2(\omega)$ is imaginary for every real ω. Therefore, in this case the terms $v_{2,+}$ and $v_{2,-}$ represent standing waves and we neglect them to get the single term solution

$$v = v_+ = \frac{1}{2\pi} \int_{-\infty}^{\infty} A_+(\omega) e^{i(k(\omega)x - \omega t)}\, d\omega.$$

Due to the boundary condition, this takes the explicit form

$$v(x, t) = \frac{1}{2\pi} \int_{-\infty}^{\infty} g^F(\omega) e^{i(k(\omega)x - \omega t)}\, d\omega \tag{4.24}$$

where g^F stands for the Fourier transform of g.

The case of the *coupled system* is a bit more difficult. Then it is necessary to incorporate the optical term $v_{2,+}$, too, because $k_2(\omega)$ is real for $|\omega| \geq \sqrt{\frac{\alpha}{\delta}}$. In this case the formula

$$v(x, t) = \frac{1}{2\pi} \int_{-\infty}^{\infty} A_+(\omega) e^{i(k(\omega)x - \omega t)}\, d\omega + \frac{1}{2\pi} \int_{|\omega| > \sqrt{\frac{\alpha}{\delta}}} A_{2,+}(\omega) e^{i(k_2(\omega)x - \omega t)}\, d\omega \tag{4.25}$$

holds with

$$
\begin{aligned}
A_{+}(\omega) &= g^F(\omega) \quad \text{for } |\omega| < \sqrt{\frac{\alpha}{\delta}}, \\
A_{+}(\omega) + A_{2,+}(\omega) &= g^F(\omega) \quad \text{for } |\omega| > \sqrt{\tfrac{\alpha}{\delta}}.
\end{aligned}
\tag{4.26}
$$

The solution is not completely determined because the amplitudes of higher frequencies have still a degree of freedom. Therefore, an additional boundary condition at $x = 0$ should be specified. This condition can be imposed either for macro-quantities (macrodisplacement, macrostress etc.) or micro-quantities. Nevertheless, in case the higher frequencies have small amplitudes, the solution can be simplified by neglecting the optical term. Then (4.25), (4.26) yield the approximate explicit formula

$$
v(x,t) \approx \frac{1}{2\pi} \int_{-\infty}^{\infty} g^F(\omega) e^{i(k(\omega)x - \omega t)}\, d\omega. \tag{4.27}
$$

Secondly, let us consider the problem with given displacement at $x = 0$, i.e.,

$$
u(0,t) = h(t) \tag{4.28}
$$

with a known function h. This problem can be treated in a similar manner. The solutions in the cases of the hierarchical equation and the coupled system read

$$
v(x,t) = \frac{1}{2\pi} \int_{-\infty}^{\infty} ik(\omega) h^F(\omega) e^{i(k(\omega)x - \omega t)}\, d\omega \tag{4.29}
$$

and (4.25), respectively, where in the latter case

$$
\begin{aligned}
\frac{A_{+}(\omega)}{ik(\omega)} &= h^F(\omega) \quad \text{for } |\omega| < \sqrt{\frac{\alpha}{\delta}}, \\
\frac{A_{+}(\omega)}{ik(\omega)} + \frac{A_{2,+}(\omega)}{ik_2(\omega)} &= h^F(\omega) \quad \text{for } |\omega| > \sqrt{\frac{\alpha}{\delta}}
\end{aligned}
\tag{4.30}
$$

and h^F denotes the Fourier transform of h. Again, if higher frequency terms have small amplitudes, the latter formula (4.25) with (4.30) implies that

$$
v(x,t) \approx \frac{1}{2\pi} \int_{-\infty}^{\infty} ik(\omega) h^F(\omega) e^{i(k(\omega)x - \omega t)}\, d\omega. \tag{4.31}
$$

Evidently, the derived formulas for v (4.24), (4.25), (4.27), (4.29) and (4.31) remain valid under the more general basic assumption that $v = 0$ for $t \le t_0$ where t_0 is some constant time value.

4.2.3 Gaussian Wave Packets

In this subsection we study right-propagating waves generated by the boundary condition (4.23) involving a periodic function with an amplitude modulation, i.e.,

$$g(t) = A\Theta(t - t_0)e^{-\frac{t^2}{4\sigma^2}} e^{-i\omega_0 t} \tag{4.32}$$

where Θ is the Heaviside function:

$$\Theta(t) = \begin{cases} 1 & \text{if } t \geq 0 \\ 0 & \text{if } t < 0, \end{cases}$$

A, σ are positive constants (the amplitude and the Gaussian dispersion of the modulation), t_0 is a sufficiently large negative time value, such that $g(t_0) \approx 0$, and ω_0 is a given frequency. The medium is assumed to be in equilibrium for $t < t_0$. In the case of the coupled system we assume that $\omega_0^2 \ll \frac{\alpha}{\delta}$ to damp the optical component.

We are going to deduce the solution in a closed form for this boundary value problem. To this end, let us transform g approximating and using the table of Fourier transforms [22]:

$$\begin{aligned} g^F(\omega) &= A \int_{t_0}^{\infty} e^{-\frac{t^2}{4\sigma^2}} e^{i(\omega-\omega_0)t}\, dt \approx A \int_{-\infty}^{\infty} e^{-\frac{t^2}{4\sigma^2}} e^{i(\omega-\omega_0)t}\, dt \\ &= 2A\sigma\sqrt{\pi} e^{-\sigma^2(\omega-\omega_0)^2}. \end{aligned}$$

Combining this formula with (4.24) and (4.27), the approximate solution of the boundary value problem is

$$v(x,t) \approx \frac{A\sigma}{\sqrt{\pi}} \int_{-\infty}^{\infty} e^{-\sigma^2(\omega-\omega_0)^2} e^{i(k(\omega)x-\omega t)}\, d\omega. \tag{4.33}$$

Since $k(\omega)$, given by (4.8) or (4.16), is a complicated function, the integral in this formula can analytically be evaluated only as an approximation. We follow the ideas due to Elmore and Heald [10]. Observing that the integrand is rapidly decreasing if $|\omega - \omega_0|$ increases, we expand $k(\omega)$ into the Taylor series around $\omega = \omega_0$. Keeping the first three terms, we get

$$k(\omega) \approx k(\omega_0) + k'(\omega_0)(\omega - \omega_0) + \frac{1}{2}k''(\omega_0)(\omega - \omega_0)^2. \tag{4.34}$$

From the definitions of the phase and group velocities we express

$$k(\omega_0) = \frac{\omega_0}{c_{ph}}, \qquad k'(\omega_0) = \frac{1}{c_g}.$$

In addition, we denote

$$d = \frac{1}{2}k''(\omega_0).$$

Thus, substituting (4.34) into (4.33) we deduce that

$$v(x,t) \approx \frac{A\sigma}{\sqrt{\pi}} e^{i\omega_0(\frac{x}{c_{ph}}-t)} \int_{-\infty}^{\infty} e^{i(\omega-\omega_0)(\frac{x}{c_g}-t)+(\omega-\omega_0)^2(ixd-\sigma^2)}\, d\omega. \tag{4.35}$$

The integral inside this expression will be computed in Sect. 4.3. The resulting formula is

$$v(x,t) \approx \frac{A\sigma}{\sqrt{\sigma^2 - ixd}} e^{-f(x,t)} e^{i\omega_0(\frac{x}{c_{ph}}-t)} \tag{4.36}$$

where

$$f(x,t) = \frac{1}{4}\left(\frac{x}{c_g} - t\right)^2 \left(\sigma^2 - ixd\right)^{-1}. \tag{4.37}$$

The formula (4.36) contains the main branch of the square root, i.e.,

$$\sqrt{z} = \sqrt{|z|}\left[\cos\left(\frac{\arg z}{2}\right) + i\sin\left(\frac{\arg z}{2}\right)\right]. \tag{4.38}$$

In practice we need the real part of the solution. The extraction of this part is again included in Sect. 4.3. Here we give only the result:

$$\begin{aligned} \operatorname{Re} v(x,t) &\approx A_1(x) e^{-\frac{1}{4\sigma_1(x)^2}(\frac{x}{c_g}-t)^2} \\ &\times \cos\left[\omega_0\left(\frac{x}{c_{ph}} - t\right) + \Phi(x) - \frac{xd}{4\sigma^2\sigma_1(x)^2}\left(\frac{x}{c_g} - t\right)^2\right], \end{aligned} \tag{4.39}$$

where

$$A_1(x) = \frac{A\sigma}{\sqrt[4]{\sigma^4 + x^2 d^2}}, \tag{4.40}$$

$$\sigma_1(x) = \frac{\sqrt{\sigma^4 + x^2 d^2}}{\sigma}, \tag{4.41}$$

$$\Phi(x) = \arctan\left(\frac{xd}{2\sigma^2}\right). \tag{4.42}$$

From these formulas we see that the propagating wave has an approximate form of a harmonic wave with Gaussian amplitude modulation. The amplitude is decreasing with increasing x (expression (4.40)) and the dispersion of the modulation in (4.41) is increasing. The harmonic part in (4.39) is influenced by the phase shift $\Phi(x)$ and a frequency changing term $\frac{xd}{4\sigma^2\sigma_1(x)^2}(\frac{x}{c_g} - t)^2$. The latter term shows the increasing frequency if t departs from $t = x/c_g$. This actually shows that the accuracy of the approximation (4.39) is sufficient close to $t = x/c_g$ where the absolute value of $\operatorname{Re} v(x,t)$ is large enough.

As we saw, the deduced second (with respect to $k(\omega)$) approximation of the wave function depends on three parameters: c_{ph}, c_g and d that are related to $k(\omega_0)$, $k'(\omega_0)$ and $k''(\omega_0)$, respectively. The parameter d affects the amplitude A_1, the modulation dispersion σ_1 and the phase shift Φ (formulas (4.40)–(4.42)).

The parameters c_{ph}, c_g and d can be measured and used to reconstruct the coefficients in the inverse problems. In particular, d can be obtained from the measurement of the amplitude, modulation dispersion or the phase shift by solving one of (4.40)–(4.42).

4.3 Proofs of Mathematical Statements

Proof of Lemma 4.1 It is convenient to use the inverse function $\omega(k)$ given by (4.6) for the proof. Then $c_{ph} = \frac{\omega(k)}{k} = \sqrt{\frac{b+\delta\gamma k^2}{1+\delta\beta k^2}}$ and

$$c_g = \omega'(k) = \left[\sqrt{\frac{b+\delta\gamma k^2}{1+\delta\beta k^2}} + \frac{\delta k^2(\gamma - b\beta)}{(1+\delta\beta k^2)^2}\sqrt{\frac{1+\delta\beta k^2}{b+\delta\gamma k^2}}\right].$$

Comparing the formulas for c_{ph} and c_g, we immediately obtain the classification of the dispersion on the basis of the sign of $b\beta - \gamma$ as presented in Lemma 4.1. The formula $k(\omega) = \frac{1}{\sqrt{b}}\omega$ in case $b\beta - \gamma = 0$ immediately follows from the formula for $k(\omega)$ (4.8). □

Proof of Lemma 4.2 To prove this lemma, we have to deduce some auxiliary formulas. Differentiating (4.16), we obtain the following relation for $c_{ph} = \frac{\omega}{k}$ and $c_g = \frac{1}{k'(\omega)}$:

$$\begin{aligned} k'(\omega) &= \frac{k(\omega)}{\omega} + \frac{Q_1(\omega) + Q_2(\omega)}{2a_0 a_1 R_1(\omega) R_2(\omega)} \\ \Rightarrow \quad \frac{c_{ph} - c_g}{c_{ph} c_g} &= \frac{Q_1(\omega) + Q_2(\omega)}{2a_0 a_1 R_1(\omega) R_2(\omega)} \end{aligned} \tag{4.43}$$

where

$$Q_1(\omega) = \frac{a_0\alpha - \vartheta}{\delta\omega^2} R_1(\omega),$$

$$Q_2(\omega) = \left(a_0 - a_1 - \frac{a_0\alpha - \vartheta}{\delta\omega^2}\right)\frac{a_0\alpha - \vartheta}{\delta\omega^2} - \frac{2a_1\vartheta}{\delta\omega^2},$$

$$R_1(\omega) = \sqrt{\left(a_0 - a_1 - \frac{a_0\alpha - \vartheta}{\delta\omega^2}\right)^2 + \frac{4a_1\vartheta}{\delta\omega^2}},$$

$$R_2(\omega) = \sqrt{\frac{a_0 + a_1 - \frac{a_0\alpha - \vartheta}{\delta\omega^2} + R_1(\omega)}{2a_0a_1}}.$$

We follow the basic physical inequalities (3.43), (3.44) and the relations $R_1(\omega)$, $R_2(\omega)$, $c_{ph}(\omega), c_g(\omega) > 0$. Then, we get from (4.43)

$$\operatorname{sign}(c_{ph} - c_g) = \operatorname{sign}\big(Q_1(\omega) + Q_2(\omega)\big) \tag{4.44}$$

and by elementary calculations we obtain

$$\operatorname{sign} Q_2(\omega) = \operatorname{sign}\left[\frac{a_0(a_0\alpha - a_1\alpha - \vartheta) - a_1\vartheta}{(a_0\alpha - \vartheta)^2} - \frac{1}{\delta\omega^2}\right]. \tag{4.45}$$

Let us compare the quantities $Q_1(\omega)$ and $Q_2(\omega)$. Squaring these quantities, subtracting, simplifying and taking the inequality $Q_1(\omega) > 0$ into account, we deduce the relation

$$\operatorname{sign}\big(Q_1(\omega) - |Q_2(\omega)|\big) = \operatorname{sign}(a_0\alpha - a_1\alpha - \vartheta). \tag{4.46}$$

Summing up, in view of the relations (4.44)–(4.46) we come to the following conclusions. Firstly, in case $a_0\alpha - a_1\alpha - \vartheta > 0$ the quantity $Q_1(\omega) + Q_2(\omega)$ is everywhere positive and hence $c_{ph} > c_g$ for any $\omega \in \mathbb{R}$. Secondly, in case $a_0\alpha - a_1\alpha - \vartheta < 0$ the quantities $Q_1(\omega) - |Q_2(\omega)|$ and $Q_2(\omega)$ are negative for any $\omega \in \mathbb{R}$. This implies that $Q_1(\omega) + Q_2(\omega)$ is negative and hence $c_{ph} < c_g$ for any $\omega \in \mathbb{R}$. Finally, if $a_0\alpha - a_1\alpha - \vartheta = 0$ then $Q_1(\omega) - |Q_2(\omega)| = 0$ and $Q_2(\omega) < 0$ for any $\omega \in \mathbb{R}$. This implies that $Q_1(\omega) + Q_2(\omega) = 0$ and hence $c_{ph} = c_g$ for any $\omega \in \mathbb{R}$. The relations $k(\omega) = \frac{1}{\sqrt{a_1}}\omega$ and $k_2(\omega) = \omega\sqrt{\frac{1}{a_0}(1 - \frac{\alpha}{\delta\omega^2})}$ in case $a_0\alpha - a_1\alpha - \vartheta = 0$ follow from the formulas (4.16) and (4.17). This proves the assertions (1)–(3) of Lemma 4.2.

To prove the assertion (4), we note that

$$k_2'(\omega) = \frac{k_2(\omega)}{\omega} + \frac{Q_1(\omega) - Q_2(\omega)}{2a_0a_1R_1(\omega)R_3(\omega)}$$
$$\Rightarrow \quad \frac{c_{ph2} - c_{g2}}{c_{ph2}c_{g2}} = \frac{Q_1(\omega) - Q_2(\omega)}{2a_0a_1R_1(\omega)R_3(\omega)} \tag{4.47}$$

where Q_1, Q_2, R_1 are defined before and

$$R_3(\omega) = \sqrt{\frac{a_0 + a_1 - \frac{a_0\alpha - \vartheta}{\delta\omega^2} - R_1(\omega)}{2a_0a_1}}.$$

From (4.47) we obtain

$$\operatorname{sign}(c_{ph2} - c_{g2}) = \operatorname{sign}\big(Q_1(\omega) - Q_2(\omega)\big). \tag{4.48}$$

By virtue of the relations (4.45), (4.46), (4.48) and $Q_1 > 0$ we see that the inequality $c_{ph2} - c_{g2} > 0$ holds independently of the sign of $a_0\alpha - a_1\alpha - \vartheta$. The assertion (4) is also proved. The proof is complete. □

Proof of formulas (4.36) *and* (4.39) Let us start with (4.36). To prove this formula, we make use of the equality [22]

$$\int_{-\infty}^{\infty} e^{-(c_1+c_2 i)^2(\tau+c_3 i)^2}\, d\tau = \frac{\sqrt{\pi}}{c_1 + c_2 i} \tag{4.49}$$

that holds for any $c_1 > 0$ and $c_2, c_3 \in \mathbb{R}$. In view of (4.37) we transform the exponent in the integral (4.35) as follows:

$$\begin{aligned} &i(\omega - \omega_0)\left(\frac{x}{c_g} - t\right) + (\omega - \omega_0)^2\left(ixd - \sigma^2\right) \\ &= -f(x,t) - \left(\sigma^2 - ixd\right)\left[\omega - \omega_0 - \frac{i(\frac{x}{c_g} - t)}{2(\sigma^2 - ixd)}\right]^2. \end{aligned}$$

Thus,

$$\begin{aligned} I &:= \int_{-\infty}^{\infty} e^{i(\omega-\omega_0)(\frac{x}{c_g}-t)+(\omega-\omega_0)^2(ixd-\sigma^2)}\, d\omega \\ &= e^{-f(x,t)} \int_{-\infty}^{\infty} e^{-(\sigma^2-ixd)[\omega-\omega_0-\frac{i(\frac{x}{c_g}-t)}{2(\sigma^2-ixd)}]^2}\, d\omega. \end{aligned}$$

Changing the variable of integration

$$\tau = \omega - \omega_0 + \frac{(\frac{x}{c_g} - t)xd}{2(\sigma^4 + x^2 d^2)}$$

yields

$$I = e^{-f(x,t)} \int_{-\infty}^{\infty} e^{-(\sigma^2-ixd)[\tau-\frac{i(\frac{x}{c_g}-t)\sigma^2}{2(\sigma^4+x^2d^2)}]^2}\, d\tau.$$

Applying (4.49) we get

$$I = \frac{\sqrt{\pi} e^{-f(x,t)}}{\sqrt{\sigma^2 - ixd}}.$$

Finally, substituting this formula into (4.35) we deduce (4.36) with (4.37).

Next, let us deduce (4.39). Using (4.38) we obtain

$$\frac{A\sigma}{\sqrt{\sigma^2 - ixd}} = \frac{A\sigma\sqrt{\sigma^2 + ixd}}{|\sigma^2 - ixd|}$$
$$= \frac{A\sigma}{\sqrt[4]{\sigma^4 + x^2d^2}}\left[\cos\arctan\frac{xd}{2\sigma^2} + i\sin\arctan\frac{xd}{2\sigma^2}\right]. \quad (4.50)$$

Moreover, in view of

$$f(x,t) = \frac{(\frac{x}{c_g} - t)^2}{4(\sigma^4 + x^2d^2)}(\sigma^2 + ixd)$$

we have

$$e^{-f(x,t)}e^{i\omega_0(\frac{x}{c_{ph}} - t)} = e^{-\frac{\sigma^2(\frac{x}{c_g} - t)^2}{4(\sigma^4 + x^2d^2)}}e^{i[\omega_0(\frac{x}{c_{ph}} - t) - \frac{xd(\frac{x}{c_g} - t)^2}{4(\sigma^4 + x^2d^2)}]}. \quad (4.51)$$

Multiplying (4.50) by (4.51) and extracting the real part we obtain (4.39).

Chapter 5
Inverse Problems for Linear Waves

5.1 Inverse Problems for Harmonic Waves

5.1.1 Hierarchical Equation

Let us consider the hierarchical equation in the linear case (4.2). Our aim is to reconstruct the triplet of coefficients b, β, γ in this equation. The simplest possibility is to use the measurements of the harmonic waves for this purpose. Since the number of unknowns is three, at least three different harmonic waves have to be measured. On the basis of such measurements we can formulate the following inverse problem.

IPh1 Given the wavenumbers k_j, $j = 1, 2, 3$, of three harmonic waves with the frequencies ω_j, such that ω_j^2, $j = 1, 2, 3$, are different, determine b, β and γ.

Evidently, instead of the wavenumbers, the wavelengths $l_j = \frac{1}{k_j}$ or the phase velocities $c_{ph,j} = \frac{\omega_j}{k_j}$ can be measured and used as the data of this problem. Various experimental techniques are available for phase velocity measurement, e.g., pulse-echo and the continuous wave resonance method [73]. Secondly, in practice the number of measured waves may be bigger than 3. Then the data of the inverse problem consist of pairs (ω_j, k_j), $j = 1, \ldots, N$, where $N > 3$. But as we will see later on, the additional measurements do not bring along complementary information for the reconstruction. They may only reduce the statistical impact of the measurement errors in the solution.

The usage of the explicit functions $k(\omega)$ and $\omega(k)$ in the solution of IPh1 is somewhat cumbersome. The simplest method follows directly from the dispersion relation (4.5). Indeed, in view of this relation, IPh1 is reduced to the following 3×3 linear system:

$$\delta \omega_j^2 k_j^2 \beta - \delta k_j^4 \gamma - k_j^2 b = -\omega_j^2, \quad j = 1, 2, 3. \tag{5.1}$$

Clearly, in case more measured pairs (ω_j, k_j) are available, the related linear system contains more than 3 equations and can be solved by means of least squares.

J. Janno, J. Engelbrecht, *Microstructured Materials: Inverse Problems*, Springer Monographs in Mathematics,
DOI 10.1007/978-3-642-21584-1_5, © Springer-Verlag Berlin Heidelberg 2011

In the study of uniqueness of the solution of IPh1 and other inverse problems in this chapter, a method of vanishing polynomial coefficients will be used. In order to demonstrate this method, we present the proof of uniqueness of IPh1 here, in the main text. The proofs of uniqueness results of more complicated inverse problems below will be shifted to Sect. 5.5.

In the uniqueness proof we distinguish the dispersive case $b\beta - \gamma \neq 0$ and the nondispersive case $b\beta - \gamma = 0$ (cf. Lemma 4.1). We start with the dispersive case. Suppose that IPh1 has two solutions: b, β, λ and $\widetilde{b}, \widetilde{\beta}, \widetilde{\lambda}$. Then, in addition to (5.1), the following system is satisfied:

$$\delta\omega_j^2 k_j^2 \widetilde{\beta} - \delta k_j^4 \widetilde{\gamma} - k_j^2 \widetilde{b} = -\omega_j^2, \quad j = 1, 2, 3. \tag{5.2}$$

Let us eliminate the quantity ω_j from these relations. To this end, we multiply (5.1) by $\delta k_j^2 \widetilde{\beta} + 1$, (5.2) by $\delta k_j^2 \beta + 1$, and subtract. Then we obtain the following equations:

$$k_j^2\big(\delta k_j^2 \widetilde{\gamma} + \widetilde{b}\big)\big(\delta k_j^2 \beta + 1\big) - k_j^2\big(\delta k_j^2 \gamma + b\big)\big(\delta k_j^2 \widetilde{\beta} + 1\big) = 0, \quad j = 1, 2, 3.$$

Evidently, $k_j \neq 0$ because $\omega_j \neq 0$. Therefore, we can divide by k_j^2 the obtained equations. The resulting relations can be rewritten in the form

$$\delta^2(\widetilde{\gamma}\beta - \gamma\widetilde{\beta})k_j^4 + \delta(\widetilde{\gamma} - \gamma + \widetilde{b}\beta - b\widetilde{\beta})k_j^2 + \widetilde{b} - b = 0, \quad j = 1, 2, 3. \tag{5.3}$$

The latter relations show that the three numbers $z = k_j^2$, $j = 1, 2, 3$, are the roots of the following quadratic function:

$$\mathcal{P}_2(z) = \delta^2(\widetilde{\gamma}\beta - \gamma\widetilde{\beta})z^2 + \delta(\widetilde{\gamma} - \gamma + \widetilde{b}\beta - b\widetilde{\beta})z + \widetilde{b} - b. \tag{5.4}$$

We note that the numbers $k_j^2 = [k(\omega_j)]^2$, $j = 1, 2, 3$, are different because the function $k(\omega)$ is strictly increasing and ω_j^2, $j = 1, 2, 3$, are different. However, a nontrivial quadratic function may have maximally 2 different roots. Thus, $\mathcal{P}_2$ must be trivial, i.e., identically zero. This implies that the coefficients of $\mathcal{P}_2$ vanish and we get the equations

$$\widetilde{\gamma}\beta - \gamma\widetilde{\beta} = 0, \qquad \widetilde{\gamma} - \gamma + \widetilde{b}\beta - b\widetilde{\beta} = 0, \qquad \widetilde{b} - b = 0. \tag{5.5}$$

The third equation in (5.5) automatically gives $\widetilde{b} = b$. This means that the second equation in (5.5) is transformed to the form

$$b(\widetilde{\beta} - \beta) - \widetilde{\gamma} + \gamma = 0. \tag{5.6}$$

In addition, the first equation in (5.5) can be rewritten as

$$\gamma(\widetilde{\beta} - \beta) - \beta(\widetilde{\gamma} - \gamma) = 0. \tag{5.7}$$

Note that (5.6) and (5.7) form a 2×2 linear homogeneous system for $\widetilde{b} - b$ and $\widetilde{\gamma} - \gamma$. The determinant of this system equals

$$\begin{vmatrix} b & -1 \\ \gamma & -\beta \end{vmatrix} = -(b\beta - \gamma)$$

and is different from zero in the dispersive case. Therefore, the solution of the system (5.6), (5.7) is trivial, i.e., $\widetilde{\beta}-\beta=\widetilde{\gamma}-\gamma=0$. This together with the previously shown relation $\widetilde{b}=b$ implies that the solution of IPh1 is unique.

In the nondispersive case $b\beta-\gamma=0$ we have $k(\omega)=\frac{1}{\sqrt{b}}\omega$ (see Lemma 4.1). Therefore,

$$b=\frac{\gamma}{\beta}=\frac{\omega_j^2}{k_j^2},\quad j=1,2,3. \tag{5.8}$$

It is not possible to separate γ and β from the quotient $\frac{\gamma}{\beta}$.

Let us summarise the obtained results in the form of a theorem.

Theorem 5.1 *The following statements are valid for* IPh1.

(i) *In the dispersive case the solution is unique.*
(ii) *In the nondispersive case infinitely many solutions occur: only b and the quotient $\frac{\gamma}{\beta}$ can be uniquely reconstructed from the data by means of the formula* (5.8).

It is very easy to establish the dispersivity during the practical solution: in the dispersive case all moduli of phase velocities $|c_{ph,j}|=|\frac{\omega_j}{k_j}|$ are different, but in the nondispersive case they are equal to each other.

5.1.2 Coupled System

Now we deal with the coupled system in the linear case (4.11), (4.12). Again, we try to reconstruct the coefficients of this system from measurements of harmonic waves in macro-level. In the present case we may use both acoustic and optical waves. The optical waves occur if $|\omega|>\sqrt{\frac{\alpha}{\delta}}$, as we saw in Sect. 4.1.2. Note that the dispersion relation (4.14) with (4.15) and its solutions (4.16), (4.17) contain the parameters ϑ_0 and ϑ_1 only in the form of the product $\vartheta=\vartheta_0\vartheta_1$. Therefore, ϑ_0 and ϑ_1 cannot be separated from measurements of such waves. We may expect to determine the quadruple $a_0, a_1, \alpha, \vartheta$. Let us pose the following inverse problem:

IPh2 Given the wavenumbers k_j, $j=1,\dots,4$, of four (acoustic or optical) harmonic waves with the frequencies ω_j, such that ω_j^2, $j=1,\dots,4$, are different, determine a_0, a_1, α and ϑ.

Again, the wavelengths $l_j=\frac{1}{k_j}$ or the phase velocities $c_{ph,j}=\frac{\omega_j}{k_j}$ can be used as the data of the inverse problem, as well.

The problem IPh2 is decomposed into two subproblems (they are also steps in the practical solution of IPh2):

(1) determine the coefficients $\varkappa_1,\dots,\varkappa_4$ of (4.14) by means of the given pairs (ω_j,k_j), $j=1,\dots,4$;

(2) solve the system (4.15) for a_0, a_1, α and ϑ by means of the computed values of $\varkappa_1, \dots, \varkappa_4$.

The first subproblem is the 4×4 linear system

$$\omega_j^2 k_j^2 \varkappa_1 + k_j^4 \varkappa_2 + \omega_j^2 \varkappa_3 + k_j^2 \varkappa_4 = -\omega_j^4, \quad j = 1, \dots, 4, \tag{5.9}$$

for the unknowns $\varkappa_1, \dots, \varkappa_4$. In case more than 4 harmonic waves are measured, the system (5.9) contains more equations and can be solved by means of least squares.

We make this decomposition of IPh2 only in the dispersive case. In the nondispersive case it is much easier to use the simple formulas of $k(\omega)$ and $k_2(\omega)$ from the item 3 of Lemma 4.2 for the solution.

In the dispersive case we study the uniqueness for the subproblems (5.9) and (4.15), respectively. For the first subproblem the following theorem holds.

Theorem 5.2 *In the dispersive case the solution of the system* (5.9) *is unique.*

The proof of this theorem is contained in Sect. 5.5.

Now let us consider the second subproblem. The first equations in (4.15) form an independent subsystem for a_0 and a_1. It has two pairs of solutions $(a_0, a_1) = (a_{0,1}, a_{1,1})$ and $(a_0, a_1) = (a_{0,2}, a_{1,2})$ where

$$\begin{aligned} a_{0,1} &= \frac{-\varkappa_1 + \sqrt{\varkappa_1^2 - 4\varkappa_2}}{2}, & a_{1,1} &= \frac{-\varkappa_1 - \sqrt{\varkappa_1^2 - 4\varkappa_2}}{2}, \\ a_{0,2} &= \frac{-\varkappa_1 - \sqrt{\varkappa_1^2 - 4\varkappa_2}}{2}, & a_{1,2} &= \frac{-\varkappa_1 + \sqrt{\varkappa_1^2 - 4\varkappa_2}}{2}. \end{aligned} \tag{5.10}$$

The third equation in (4.15) gives $\alpha = -\delta\varkappa_3$. From the fourth equation in (4.15) we get the formula for ϑ, namely $\vartheta = a_0\alpha - \delta\varkappa_4$. The quantity ϑ depends on the chosen value of a_0. Consequently, the second subproblem has two solutions:

$$a_0 = a_{0,1}, \qquad a_1 = a_{1,1}, \qquad \alpha = -\delta\varkappa_3, \qquad \vartheta = \vartheta_1 := a_{0,1}\alpha + \delta\varkappa_4, \tag{5.11}$$

$$a_0 = a_{0,2}, \qquad a_1 = a_{1,2}, \qquad \alpha = -\delta\varkappa_3, \qquad \vartheta = \vartheta_2 := a_{0,2}\alpha + \delta\varkappa_4. \tag{5.12}$$

Let us select the solutions that meet the physical restrictions (3.43) and (3.44). In view of the definitions of ϑ_1 and ϑ_2 the relation

$$\vartheta_2 = a_{0,2}\alpha - a_{0,1}\alpha + \vartheta_1 \tag{5.13}$$

holds. Since $a_{0,2} \le a_{0,1}$ (see (5.10)) and $\alpha > 0$ we have from (5.13)

$$\vartheta_1 \ge \vartheta_2.$$

Consequently, two different cases may occur:

$$\text{either} \quad \vartheta_1 > 0, \ \vartheta_2 \le 0 \quad \text{or} \quad \vartheta_1 > 0, \ \vartheta_2 > 0.$$

The third case $\vartheta \le 0$, $\vartheta_2 \le 0$ is impossible because then neither of the solutions (5.11) and (5.12) meets the physical condition $\vartheta > 0$.

In the case $\vartheta_1 > 0$, $\vartheta_2 \le 0$ only the first solution (5.11) is physical. Then due to (5.13) and the relation $a_{0,2} = a_{1,1}$ (see (5.10)) we have

$$0 \ge \vartheta_2 = a_{0,2}\alpha - a_{0,1}\alpha + \vartheta_1 = a_{1,1}\alpha - a_{0,1}\alpha + \vartheta_1.$$

This implies that $a_{0,1}\alpha - a_{1,1}\alpha - \vartheta_1 \ge 0$. Thus, by virtue of the dispersivity assumption $a_0\alpha - a_1\alpha - \vartheta \ne 0$, we see that the material has the normal dispersion (cf. Lemma 4.2 for the types of dispersion).

In the case $\vartheta_1 > 0$, $\vartheta_2 > 0$ both solutions (5.11) and (5.12) are physical. Again, in view of (5.13) and the relations $a_{0,2} = a_{1,1}$, $a_{0,1} = a_{1,2}$ we obtain

$$0 < \vartheta_2 = a_{0,2}\alpha - a_{0,1}\alpha + \vartheta_1 = a_{1,1}\alpha - a_{0,1}\alpha + \vartheta_1,$$

$$0 < \vartheta_1 = a_{0,1}\alpha - a_{0,2}\alpha + \vartheta_2 = a_{1,2}\alpha - a_{0,2}\alpha + \vartheta_2.$$

This yields $a_{0,1}\alpha - a_{1,1}\alpha - \vartheta_1 < 0$ and $a_{0,2}\alpha - a_{1,2}\alpha - \vartheta_2 < 0$. Therefore, the material has the anomalous dispersion.

Let us compare the solutions (5.11) and (5.12). The component α is the same. Moreover, if $\varkappa_1^2 - 4\varkappa_2 > 0$ then by (5.10) $a_{0,1} > a_{0,2}$ and hence the components a_0, a_1 and ϑ of the solutions are different. But in case $\varkappa_1^2 - 4\varkappa_2 = 0$ the solutions (5.11) and (5.12) entirely coincide. Then $a_0 = a_{0,j} = a_{1,j} = a_1$, i.e., this is the midpoint of the anomalous dispersion.

Let us summarise the obtained results.

Lemma 5.1 *The following statements are valid for the second subproblem.*

(i) *In the case of the normal dispersion the solution is unique and has the form* (5.11).
(ii) *In the case of the anomalous dispersion two solutions occur: they are given by* (5.11) *and* (5.12)*, contain the same value of α, but the other components a_0, a_1 and ϑ coincide only in the case of the midpoint of the anomaly.*

Putting Theorem 5.2 and Lemma 5.1 together, we have the next theorem.

Theorem 5.3 *The following statements are valid for* IPh2.

(i) *In the case of the normal dispersion the solution is unique.*
(ii) *In the case of the anomalous dispersion two solutions occur: they contain the same value of α, but the other components a_0, a_1 and ϑ of the solutions coincide only in the case of the midpoint of the anomaly.*

It remains to consider the nondispersive case when $a_0\alpha - a_1\alpha - \vartheta = 0$. Due to the assertion 3 of Lemma 4.2 any measurement pair (ω_j, k_j) gathered from an acoustic harmonic wave determines uniquely the parameter a_1:

$$a_1 = \frac{\omega_j^2}{k_j^2}. \tag{5.14}$$

But any two measurement pairs (ω_{j_i}, k_{j_i}), $i = 1, 2$, from optical harmonic waves give the following linear system for a_0 and α:

$$k_{j_i}^2 a_0 + \frac{1}{\delta}\alpha = \omega_{j_i}^2, \quad i = 1, 2. \tag{5.15}$$

The matrix of this system is regular because $k_{j_1}^2 \neq k_{j_2}^2$. This follows from the assumption $\omega_{j_1}^2 \neq \omega_{j_2}^2$ and the strict monotonicity of $k_2(\omega)$. Therefore, the solution of (5.15) is unique. The parameter ϑ is given in terms a_1, a_0 and α by $\vartheta = a_0\alpha - a_1\alpha$. Summing up, the determination of the full vector of coefficients is possible in case

$$\begin{gathered}\text{the set of frequency-wavenumber pairs } \{(\omega_j, k_j),\ j = 1, \ldots, 4\} \\ \text{contains at least one pair from an acoustic wave and} \\ \text{at least two pairs from optical waves}\end{gathered} \tag{5.16}$$

and we can formulate the next result.

Theorem 5.4 *The following statements are valid for* IPh2 *in the nondispersive case.*

(i) *If* (5.16) *holds then the solution is unique.*
(ii) *In the opposite case infinitely many solutions occur.*

Again, it is possible to determine whether the solution is unique or non-unique during the practical solving procedure. Firstly, if the data of IPh2 have the property (5.16) then the system (5.9) is always regular and the first step can be successfully performed to get the quantities $\varkappa_1, \ldots, \varkappa_4$. The second step is to be started with the computation of the first solution by the formula (5.11). In case this solution satisfies the condition $a_0\alpha - a_1\alpha - \vartheta \geq 0$, it is the unique solution of the inverse problem. But in case the condition $a_0\alpha - a_1\alpha - \vartheta < 0$ holds, the second solution exists too, and is to be computed by (5.12). These two solutions coincide when $\varkappa_1^2 - 4\varkappa_2 = 0$.

Secondly, if the property (5.16) is not valid then the system (5.9) may be singular or regular. In the singular case the material is nondispersive and IPh2 has infinitely many solutions that can be partially recovered by means of the formulas (5.14), (5.15) and $\vartheta = a_0\alpha - a_1\alpha$ depending on given data. But in case the system (5.9) is regular, it provides unique quantities $\varkappa_1, \ldots, \varkappa_4$ and the second step can be completed as above.

Finally, we remark that it is possible to separate ϑ_0 and ϑ_1 from the product $\vartheta = \vartheta_0\vartheta_1$ in case certain information about the microdeformation is available, too. More precisely, let us be given the amplitudes of the macro- and microdeformation A and $\widehat{A}$, respectively, of the first wave with $\omega = \omega_1$ and $k = k_1$. Then, due to (4.13) we get the parameter ϑ_0 by the formula

$$\vartheta_0 = \frac{A(\omega_1^2 - a_0 k_1^2)}{\widehat{A} k_1^2}. \tag{5.17}$$

5.1.3 General Consequences

Let us make some general conclusions from the results of the previous two subsections. It is natural to ask: is it possible to improve the nonuniqueness results of Theorems 5.1, 5.3 and 5.4 if we incorporate measurements of more harmonic waves or superpositions of such waves? Clearly, the nonuniqueness assertion Theorem 5.1(ii) remains valid in such generalisations. Every harmonic component of a nondispersive wave is governed by the simple linear relation $k = \frac{1}{\sqrt{b}}\omega$ and hence contains information about b only. Therefore, the following statement holds.

Corollary 5.1 *In the nondispersive case any superposition of harmonic wave solutions of the hierarchical equation does not contain enough information to recover all parameters: it is not possible to determine more than* $b = \frac{\beta}{\gamma}$.

Further, we ask: can we improve the assertion (ii) of Theorem 5.3 if we provide more measurements of harmonic wave packets? Again, the answer is no. The nonuniqueness in this theorem is caused by the properties of the nonlinear system (4.15) that is to be solved in the second step of the solution. Incorporating more harmonic components only overdetermines the system (5.9) which is to be solved in the first step and whose solution is already unique. Thus, we may formulate the following statement.

Corollary 5.2 *In the cases of weak and strong anomalous dispersion any superposition of harmonic wave solutions of the coupled system in macro-level does not contain enough information to recover all parameters: it is not possible to determine more than a single value for* α *and two different values for the other parameters.*

Theorem 5.4 shows that nondispersive waves in the coupled system contain enough information to recover uniquely all parameters only in case they contain at least one acoustic component and two different optical components. In the opposite case those waves are not sufficiently informative. For instance, concerning the acoustic waves packets the following statement holds.

Corollary 5.3 *In the nondispersive case any superposition of acoustic harmonic wave solutions of the coupled system in macro-level does not contain enough information to recover all parameters: it is not possible to determine more than* $a_1 = a_0 - \frac{\vartheta}{\alpha}$.

5.2 Inverse Problems for Gaussian Wave Packets

In this section we discuss the reconstruction of parameters from measurements of Gaussian wave packets. We focus ourselves on some problems that make use of the phase and group velocities and in some cases the dispersion parameter $d = \frac{k''(\omega)}{2}$, too. The parameter d can be extracted from measurements of the amplitude change and modulation dispersion of the phase shift solving one of (4.40)–(4.42). Probably it is most realistic to measure the amplitude change. However, the amplitude or modulation dispersion provide d^2, hence leave the sign of d open. The sign of d may also be determined from the additional observation of the sign of the phase shift, namely

$$\operatorname{sign} d = \operatorname{sign} \Phi(x).$$

Firstly, we pose and study some inverse problems for the hierarchical equation.

IPg1 Given the phase velocity c_{ph}, the group velocity c_g and d of a single wave packet with the central frequency ω_0, determine b, β and γ.

The data of this problem are related to first and second order derivatives of the dispersion function. From the basic dispersion equation (4.5), by differentiation we immediately deduce the following equations for $k' = k'(\omega)$ and $k'' = k''(\omega)$:

$$\delta\beta\omega_0 k\big(k + \omega k'\big) - 2\delta\gamma k^3 k' + \omega - bkk' = 0,$$

$$\delta\beta\big[k^2 + 4\omega kk' + \omega^2 (k')^2 + \omega^2 kk''\big] - 2\delta\gamma k^2\big[3(k')^2 + kk''\big] + 1 - b\big[(k')^2 + kk'\big] = 0.$$

Therefore, IPg1 is equivalent to the following 3×3 linear system for b, β, γ:

$$\left.\begin{aligned} &\delta\omega_0^2 k_0^2 \beta - \delta k_0^4 \gamma - k_0^2 b = -\omega_0^2, \\ &\delta\omega_0 k_0 \big(k_0 + \omega_0 k_0'\big)\beta - 2\delta k_0^3 k_0' \gamma - k_0 k_0' b = -\omega_0, \\ &\delta\big[k_0^2 + 4\omega_0 k_0 k_0' + \omega_0^2 (k_0')^2 + \omega_0^2 k_0 k_0''\big]\beta - 2\delta k_0^2\big[3(k_0')^2 + k_0 k_0''\big]\gamma \\ &\quad - \big[(k_0')^2 + k_0 k_0''\big] b = -1, \end{aligned}\right\} \tag{5.18}$$

where $k_0 = \frac{\omega_0}{c_{ph}}$, $k_0' = \frac{1}{c_g}$ and $k_0'' = 2d$.

Another reconstruction procedure uses only phase and group velocities of the wave packets. In such a case at least two wave packets are to be incorporated. Let us pose the related inverse problem.

IPg2 Given the phase and group velocities $c_{ph,1}$ and $c_{g,1}$ of the first wave packet with the central frequency ω_1, and the phase velocity $c_{ph,2}$ of the second wave packet with the central frequency ω_2, such that $\omega_1^2 \neq \omega_2^2$, determine b, β and γ.

From the dispersion equation (4.5) and the corresponding equation for $k' = k'(\omega)$ we infer the following 3×3 linear system that is equivalent to IPg2:

$$\left.\begin{aligned} &\delta\omega_j^2 k_j^2 \beta - \delta k_j^4 \gamma - k_j^2 b = -\omega_j^2, \quad j = 1, 2, \\ &\delta\omega_1 k_1 \big(k_1 + \omega_1 k_1'\big)\beta - 2\delta k_1^3 k_1' \gamma - k_1 k_1' b = -\omega_1. \end{aligned}\right\} \tag{5.19}$$

Here $k_j = \frac{\omega_j}{c_{ph,j}}$, $j = 1, 2$, and $k_1' = \frac{1}{c_{g,1}}$.

Theorem 5.5 *The following statements are valid for* IPg1 *and* IPg2.

(i) *In the dispersive case the solution is unique.*
(ii) *In the nondispersive case infinitely many solutions occur: only b and the quotient $\frac{\gamma}{\beta}$ can be uniquely reconstructed from the data by the formula*

$$b = \frac{\gamma}{\beta} = c_g^2. \tag{5.20}$$

The proof can be found in Sect. 5.5.

Clearly, in practice more information may be available, for instance, the phase and group velocities and the parameters d of several Gaussian wave packets. The formulated inverse problems incorporate minimum amounts of information sufficient for the unique reconstruction in the dispersive case.

Secondly, let us consider the determination of the parameters of the coupled system. We pose the following problem.

IPg3 Given the phase and group velocities $c_{ph,j}, c_{g,j}$, $j = 1,2$, of two wave packets with the central frequencies ω_j, such that $\omega_1^2 \neq \omega_2^2$, determine a_0, a_1, α and ϑ.

The structure of this problem is similar to the structure of the inverse problem for harmonic waves IPh2 studied in Sect. 5.1.2. Namely, we decompose IPg3 into two subproblems:

(1) determine the coefficients $\varkappa_1, \ldots, \varkappa_4$ of (4.14) by means of the data $c_{ph,j}, c_{g,j}$, $j = 1,2$;
(2) solve the system (4.15) for a_0, a_1, α and ϑ by means of the computed values of $\varkappa_1, \ldots, \varkappa_4$.

The first subproblem is again equivalent to a linear system. Indeed, let us differentiate (4.14) with respect to ω:

$$2\omega^3 + \varkappa_1\left(\omega k^2 + \omega^2 k k'\right) + 2\varkappa_2 k^3 k' + \varkappa_3 \omega + \varkappa_4 k k' = 0.$$

Observing this expression and (4.14) we see that $\varkappa_1, \ldots, \varkappa_4$ is the solution vector of the following 4×4 system:

$$\left.\begin{aligned} & k_j^2\omega_j^2\varkappa_1 + k_j^4\varkappa_2 + \omega_j^2\varkappa_3 + k_j^2\varkappa_4 = -\omega_j^4, \quad j = 1,2, \\ & \left(\omega_j k_j^2 + \omega_j^2 k_j k_j'\right)\varkappa_1 + 2k_j^3 k_j'\varkappa_2 + \omega_j\varkappa_3 + k_j k_j'\varkappa_4 \\ & \quad = -2\omega_j^3, \quad j = 1,2. \end{aligned}\right\} \tag{5.21}$$

Theorem 5.6 *In the dispersive case the solution of the system* (5.21) *is unique.*

The proof is contained in Sect. 5.5.

To formulate a uniqueness result concerning IPg3, we combine Theorem 5.6 and Lemma 5.1 in the dispersive case and apply Corollary 5.3 in the nondispersive case. In the latter situation it is possible to use the formula $\frac{1}{c_g} = k'(\omega) = \frac{1}{\sqrt{a_1}}$ following from Lemma 4.2.

Theorem 5.7 *The following statements are valid for* IPg3.

(i) *In the case of the normal dispersion the solution is unique.*
(ii) *In the case of the anomalous dispersion two solutions occur: they contain the same value of* α*, but the other components* a_0*,* a_1 *and* ϑ *of the solutions coincide only in the case of the midpoint of the anomaly.*
(iii) *In the nondispersive case infinitely many solutions occur: only* a_1 *and the quantity* $a_0 - \frac{\vartheta}{\alpha}$ *can be uniquely reconstructed from the data by the formula*

$$a_1 = a_0 - \frac{\vartheta}{\alpha} = c_g^2. \tag{5.22}$$

Finally, we mention that it is possible to pose and study inverse problems for the coupled system that involves the quantity d in the data set. But those problems are technically very complicated and require long computations in the proofs. Therefore, we do not present them in this book.

5.3 Reconstruction of Parameters from Spectra of Waves

It is possible to reconstruct the parameters in our models by means of more complex linear waves, too. The idea consists in extracting harmonic counterparts from the spectral decomposition of the wave, and reducing the problem to the inverse problem for harmonic waves discussed in Sect. 5.1. In practice, this means the determination of frequency-wavenumber pairs (ω_m, k_m) from the spectra and solution of either IPh1 or IPh2.

5.3.1 The Case of Deformation Boundary Condition

Let us consider right-propagating waves on the half-line $x > 0$ generated by the deformation boundary condition (4.23). As we saw in Sect. 4.2.2, the wave function is given by (4.24) and (4.25) with (4.26) in the cases of the hierarchical equation and the coupled system, respectively. This means that the Fourier transform of the wave function possesses the formula

$$v^F(x,\omega) = g^F(\omega)e^{ik(\omega)x}$$

(in the case of the coupled system this formula holds for lower frequencies $|\omega| < \sqrt{\frac{\alpha}{\delta}}$).

Suppose that the deformation function $v(x,t)$ is measured at some point $x_1 > 0$ over the time t. Upon computation of the Fourier transforms of the data, the equation

$$e^{ik(\omega)x_1} = \frac{v^F(x_1,\omega)}{g^F(\omega)} \tag{5.23}$$

can be solved for the function $k(\omega)$. By means of this function the frequency-wavenumber pairs (ω_m, k_m) for IPh1 or IPh2 can be computed.

At first sight, the solution of (5.23) is complicated because of the periodicity of the outer component e^{iz}. Nevertheless, the right solution can easily be extracted observing the qualitative behaviour of $e^{ik(\omega)x_1}$ over some frequency interval.

Let us make use of the real part of the quotient $\frac{v^F(x_1,\omega)}{g^F(\omega)}$ during the solution. Then we have to solve the equation

$$\cos\big[k(\omega)x_1\big] = \Re\frac{v^F(x_1,\omega)}{g^F(\omega)} \tag{5.24}$$

for $k(\omega)$. It is necessary to invert the cosine in a proper way. Let us think as follows. Since $k(\omega)$ is strictly increasing and the relations $k(0) = 0$, $\lim_{w\to\infty} k(\omega) = \infty$ hold, the function $\cos[k(\omega)x_1]$ oscillates between 1 and -1. More precisely, $\cos[k(\omega)x_1]$ decreases for $\omega \in \mathcal{I}_0 = (0, \zeta_1)$, increases for $\omega \in \mathcal{I}_1 = (\zeta_1, \zeta_2)$ and so on, where

$$0 < \zeta_1 < \zeta_2 < \cdots$$

is some increasing sequence of real numbers. Thus, for the right inversion of the cosine it is necessary to find the intervals of monotonicity $\mathcal{I}_j$ of the known right-hand side $\Re\frac{v^F(x_1,\omega)}{g^F(\omega)}$. Then the desired function $k(\omega)$ can be evaluated by the formula

$$k(\omega) = \frac{1}{x_1}\left[(-1)^n \arccos\Re\frac{v^F(x_1,\omega)}{g^F(\omega)} + \pi(n+\theta_n)\right] \quad \text{for } \omega \in \mathcal{I}_n. \tag{5.25}$$

Here $\theta_n = 0$ for even n and $\theta_n = 1$ for odd n.

Let's see how this method can be practically performed. Actually, we have at our disposal a time series of measured deformations

$$v_l = v(x_1, t_l), \quad l = 1, \ldots, N$$

in an interval $[T, T_1]$, where $t_l = T + l\eta$ and $\eta = \frac{T_1 - T}{N}$. To compute the Fourier transforms, different methods may be used. Let us choose the rectangular rule for truncated Fourier integrals, because this is compatible with the standard Discrete Fourier Transform available in mathematical softwares. Then the discrete spectra of the data are

$$g^F(\omega_m) \approx \hat{g}_m = \frac{e^{iT\omega_m}}{N}\sum_{l=1}^{N} e^{\frac{2\pi ib(l-1)(m-1)}{N}} g_l,$$

$$v^F(x_1,\omega_m) \approx \hat{v}_m = \frac{e^{iT\omega_m}}{N}\sum_{l=1}^{N} e^{\frac{2\pi ib(l-1)(m-1)}{N}} v_l$$

for $m = 1, \ldots, N$, where $\tau > 0$ is the stepsize in the frequency domain, $\omega_m = (m-1)\tau$, $g_l = g((l-1)\eta)$ and $b = \frac{\eta\tau}{2\pi}$. The discrete spectrum provides the oscillat-

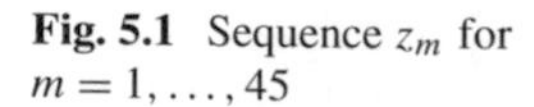

Fig. 5.1 Sequence z_m for $m = 1, \ldots, 45$

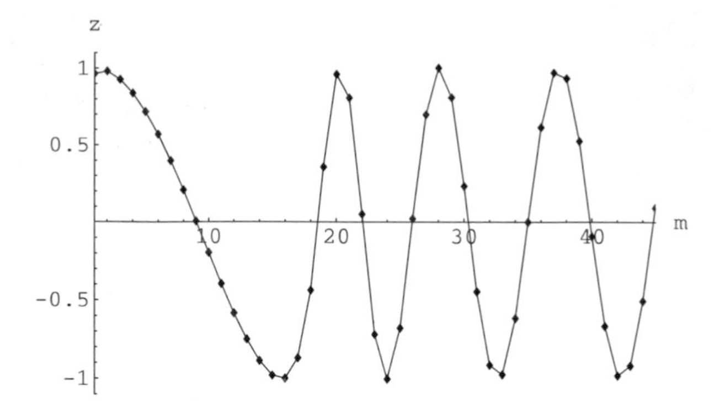

Fig. 5.2 Function $\cos[k(\omega)x_1]$ for $x_1 = 10$

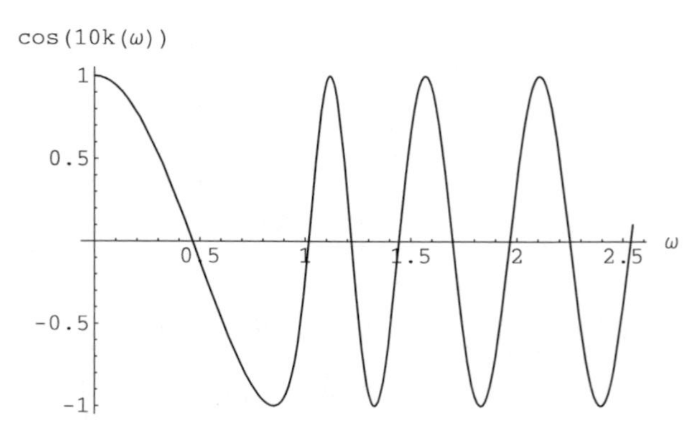

ing sequence $z_m = \mathrm{Re}\,\frac{\hat{v}_m}{\hat{g}_m}$ of real numbers that decreases for $s_0 < m < s_1$, increases for $s_1 < m < s_2$ and so on, where

$$1 = s_0 < s_1 < s_2 < \cdots$$

are some numbers. To find frequency-wavenumber pairs, it is necessary to determine the critical numbers $s_1, s_2, \ldots$. Then the wavenumber k_m corresponding to the frequency ω_m can be evaluated by means of the following formula deduced from (5.25):

$$k_m = k(\omega_m) = \frac{1}{x_1}\big[(-1)^n \arccos z_m + \pi(n + \theta_n)\big] \quad \text{for } s_n < m < s_{n+1}. \tag{5.26}$$

The formula (5.26) is applicable for all ω_m except for the critical frequencies ω_{s_n}, because the discrete problem does not contain information about the intervals of monotonicity to which ω_{s_n} belong.

For example, Fig. 5.2 shows the periodic function $\cos(k(\omega)x_1)$ corresponding to the parameters $b = 10$, $\beta = \gamma = 10^4$, $\delta = 10^{-4}$ of the hierarchical equation and $x_1 = 10$. The intervals of monotonicity are $\mathcal{I}_1 = (0, 0.87)$, $\mathcal{I}_2 = (0.87, 1.12)$, $\mathcal{I}_3 = (1.12, 1.29), \ldots$. On the top picture Fig. 5.1 the real part of the ratio of spectra $z_m = \Re \frac{\hat{v}_m}{\hat{g}_m}$ is depicted for the discrete frequencies $\omega_m = (m-1)\tau$ where $\tau = \frac{\pi}{55}$.

The latter data were computed by the standard Discrete Fourier Transform applied to the solution corresponding to the boundary excitation $g(t) = e^{-\frac{t^2}{4}}$ at $x = 0$.

The sequence z_m is oscillating with critical numbers $s_0 = 0, s_1 = 16, s_2 = 20, s_3 = 24, \ldots$. The frequency-wavenumber pairs can be obtained from the sequence z_m by means of the formula (5.26). For instance, the wavenumbers corresponding to $\omega_m = (m-1)\tau$, $m = 2, \ldots, 15$, are $k_m = \frac{1}{10}\arccos z_m$, the wavenumbers corresponding to $\omega_m = (m-1)\tau$, $m = 17, \ldots, 19$, are $k_m = \frac{1}{10}[2\pi - \arccos z_m]$ and so on.

5.3.2 *The Case of Displacement Boundary Condition*

Now we consider right-propagating waves on the half-line $x > 0$ generated by the displacement boundary condition (4.28). Then the wave function is given by (4.29) and (4.25) with (4.30) in the cases of the hierarchical equation and the coupled system, respectively. The Fourier transform of the wave function has the from

$$v^F(x, \omega) = ik(\omega)h^F(\omega)e^{ik(\omega)x}$$

(in the case of coupled system this holds for lower frequencies $|\omega| < \sqrt{\frac{\alpha}{\delta}}$).

Again, let the deformation function $v(x, t)$ be measured at some point $x_1 > 0$ over the time t. Having the Fourier transforms of the data, the equation

$$ik(\omega)e^{ik(\omega)x_1} = \frac{v^F(x_1, \omega)}{h^F(\omega)} \tag{5.27}$$

is to be solved for the function $k(\omega)$.

Note that in the present case the different physical quantities are given and measured (given displacement versus measured deformation). In turns out that such a feature reduces the periodicity problem. Indeed, taking the modulus of (5.27), we have

$$|k(\omega)| = \left|\frac{v^F(x_1, \omega)}{h^F(\omega)}\right|.$$

This means that for any positive frequency ω the corresponding wavenumber can be computed as $k = |\frac{v^F(x_1,\omega)}{h^F(\omega)}|$.

In the case of the discrete data

$$h^F(\omega_m) \approx \hat{h}_m = \frac{e^{iT\omega_m}}{N}\sum_{l=1}^{N} e^{\frac{2\pi ib(l-1)(m-1)}{N}} h_l,$$

$$v^F(x_1, \omega_m) \approx \hat{v}_m = \frac{e^{iT\omega_m}}{N}\sum_{l=1}^{N} e^{\frac{2\pi ib(l-1)(m-1)}{N}} v_l$$

with $m = 1, \ldots, N$, where $\tau > 0$ is the stepsize in the frequency domain, $\omega_m = (m-1)\tau$, $h_l = h((l-1)\eta)$, $v_l = v(x_1, (l-1)\eta)$ and $b = \frac{\eta\tau}{2\pi}$, the wavenumber k_m that corresponds to the frequency ω_m is obtained by the formula

$$k_m = \left| \frac{\hat{v}_m}{\hat{h}_m} \right|.$$

5.4 Stability and Examples

5.4.1 Stability of Solutions

Now we ask the question: under what conditions are the solutions of the studied inverse problems stable, i.e., the errors of the solutions converge to zero provided that the errors of the data tend to zero? As we saw, our inverse problems are connected with certain linear system of algebraic equations. Namely, IPh1, IPg1, IPg2 and the first subproblems of IPh2 and IPg3 are equivalent to the linear systems (5.1), (5.18), (5.19), (5.9) and (5.21), respectively.

It is well-known that the stability of a solution of a linear system of algebraic equations automatically follows from the regularity of this system, i.e., the uniqueness of the solution. (The stability in the sense of convergence of sets can be considered for singular linear systems, too, but we omit such more complicated cases here.) Therefore, due to Theorems 5.1, 5.2, 5.5 and 5.6, the solutions of IPh1, IPg1, IPg2 and the first subproblems of IPh2 and IPg3 are stable in the dispersive case.

Further, the solutions of the second subproblem of IPh2 and IPg3 are given by the explicit formulas (5.10), (5.11) and (5.12) that contain continuous functions of $\varkappa_1 \ldots, \varkappa_4$. By the continuity, the stability holds for these subproblems, too.

Summing up, we can formulate the following theorem.

Theorem 5.8 *The following statements are valid in the dispersive case.*

(i) *The unique solutions of* IPh1, IPg1 *and* IPg2 *are stable.*
(ii) *The solutions of* IPh2 *and* IPg3 (*one or two, depending on the type of the dispersion*) *are stable.*

5.4.2 Numerical Examples

We have tested the methods proposed in this chapter from the point of view of sensitivity with respect to the noise of the data. For both models (i.e. the hierarchical equation and the coupled system) the parameters were determined from the spectral composition of right-propagating waves corresponding to deformation boundary condition and Gaussian wave packets. As described above, the former problems contain as a sub-step the solution of inverse problems for harmonic waves.

Table 5.1 Relative errors in the spectral method for the hierarchical equation

ϵ	$\lvert\frac{b^\epsilon - b}{b}\rvert$	$\lvert\frac{\beta^\epsilon - \beta}{\beta}\rvert$	$\lvert\frac{\gamma^\epsilon - \gamma}{\gamma}\rvert$
0.01%	0.003%	0.011%	0.010%
0.1%	0.014%	0.12%	0.16%
1%	0.78%	2.6%	2.6%

Table 5.2 Relative errors in the spectral method for the coupled system

ϵ	$\lvert\frac{a_0^\epsilon - a_0}{a_0}\rvert$	$\lvert\frac{a_1^\epsilon - a_1}{a_1}\rvert$	$\lvert\frac{\alpha^\epsilon - \alpha}{\alpha}\rvert$	$\lvert\frac{\vartheta^\epsilon - \vartheta}{\vartheta}\rvert$
0.01%	0.072%	0.051%	0.058%	0.61%
0.1%	0.60%	0.47%	0.62%	5.2%
1%	3.6%	2.2%	3.3%	35%

The basic parameter choice for the coupled system was $a_0 = 100$, $a_1 = 1$, $\alpha = 10^{-4}$, $\vartheta = 0.002$ (the parameters α and ϑ contain the small quantity l^2, and hence it is natural to take them small). Then the corresponding parameters of the hierarchical equation are $b = 80$, $\beta = \gamma = 2 \times 10^5$ (computed by (3.45)). In all examples we took $\delta = 10^{-4}$.

The relative noise level of the data is denoted by ϵ and the computed parameters containing the noise are denoted by $a_0^\epsilon, a_1^\epsilon, \alpha^\epsilon, \vartheta^\epsilon$ (coupled system) and $b^\epsilon, \beta^\epsilon, \gamma^\epsilon$ (hierarchical equation).

For the method of spectral decomposition the boundary impulse $g(t) = e^{\frac{-t^2}{4}}$ at $x = 0$ was chosen and the solution $v(x, t)$ corresponding to prescribed (or exact) parameters computed at $x_1 = 10$ for $t \in [0, 50]$. This solution was perturbed as follows

$$v^\epsilon(x_1, t_j) = v(x_1, t_j)(1 + R_j \epsilon)$$

where $t_j = j\tau$ are discrete time values with the step $\tau = 0.01$ and R_j is the uniformly distributed random number in the interval $[-1, 1]$. The time series $v^\epsilon(x_1, t_j)$ was used as the synthetic data for the reconstruction procedure.

All computations were repeated 100 times with different random vectors R_j and the biggest relative errors selected. Tables 5.1 and 5.2 show the relative errors in the hierarchical equation and coupled system, respectively.

Another reconstruction method consists in the usage of phase and group velocities c_{ph}, c_g and in IPg1 also the dispersion quantity d of Gaussian wave packets. We chose the packets with the initial amplitude $A = 100$, the Gaussian dispersion $\sigma = 0.1$ and the following central frequencies: $\omega_0 = 300$ for problem IPg1 and $\omega_1 = 100$, $\omega_2 = 400$ for problems IPg2 and IPg3. The exact velocities were computed by the formulas $c_{ph,j} = \frac{\omega_j}{k(\omega_j)}$, $c_{q,j} = \frac{1}{k'(\omega_j)}$, $j \in \{0; 1; 2\}$, where $k(\omega)$ is given by (4.8) or (4.16), and perturbed in the following manner:

$$c_{ph,j}^\epsilon = c_{ph,j}\big(1 + R_{ph}^j \epsilon\big), \qquad c_{g,j}^\epsilon = c_{g,j}\big(1 + R_g^j \epsilon\big)$$

Table 5.3 Relative errors in IPg1

ϵ	$\lvert\frac{b^\epsilon-b}{b}\rvert$	$\lvert\frac{\beta^\epsilon-\beta}{\beta}\rvert$	$\lvert\frac{\gamma^\epsilon-\gamma}{\gamma}\rvert$
0.01%	0.029%	0.037%	0.073%
0.1%	0.072%	0.095%	0.51%
1%	2.2%	3.4%	6.2%

Table 5.4 Relative errors in IPg2

ϵ	$\lvert\frac{b^\epsilon-b}{b}\rvert$	$\lvert\frac{\beta^\epsilon-\beta}{\beta}\rvert$	$\lvert\frac{\gamma^\epsilon-\gamma}{\gamma}\rvert$
0.01%	0.004%	0.007%	0.017%
0.1%	0.022%	0.045%	0.23%
1%	1.1%	2.1%	3.3%

Table 5.5 Relative errors in IPg3

ϵ	$\lvert\frac{a_0^\epsilon-a_0}{a_0}\rvert$	$\lvert\frac{a_1^\epsilon-a_1}{a_1}\rvert$	$\lvert\frac{\alpha^\epsilon-\alpha}{\alpha}\rvert$	$\lvert\frac{\vartheta^\epsilon-\vartheta}{\vartheta}\rvert$
0.01%	0.084%	0.008%	0.024%	0.46%
0.1%	0.81%	0.034%	0.19%	5.3%
1%	7.3%	0.86%	1.2%	73%

where R_{ph}^j and R_g^j are uniformly distributed random numbers in the interval $[-1, 1]$.

The quantity d involved in IPg1 is not directly measurable. But it is possible to compute it by means of the measurable amplitude change $A_1(x)$ at some point x, making use of the following formula deduced from (4.40):

$$d = \operatorname{sign} d \frac{\sigma^2}{x} \sqrt{\frac{A^4}{A_1^4(x)} - 1} \tag{5.28}$$

where $\operatorname{sign} d = \operatorname{sign} \Phi(x)$. A synthetic datum d was constructed in the following manner. Exact d and $A_1 = A_1(x)$ at $x = 10$ were evaluated by the formulas $d = \frac{k''(\omega_0)}{2}$ and (4.40). Thereupon A_1 was perturbed:

$$A_1^\epsilon = A_1\left(1 + R_{A_1}^j \epsilon\right)$$

where $R_{A_1}^j$ are again uniformly distributed random numbers in the interval $[-1, 1]$. Finally the perturbed d^ϵ was computed by inserting A_1^ϵ into (5.28).

Summing up, the quantities $c_{ph,j}^\epsilon, c_{g,j}^\epsilon$ and in IPg1 also d^ϵ formed the synthetic data for the inverse problems. The results for IPg1, IPg2 and IPg3 are presented in Tables 5.3, 5.4, 5.5.

First of all, the numerical results support the theoretical statements about the asymptotical stability: if ϵ tends to zero then the errors of the components of the solutions also approach zero.

The computations show that the inverse problems for the linear hierarchical equation are less sensitive with respect to the noise of the data than the inverse problems for the linear coupled system. The cause is that the condition number of the matrices of these problems is amplified by the increase of the dimension: from 3 in the hierarchical equation to 4 in the coupled system. Worst results are obtained for ϑ. But one cannot make a conclusion that the reconstruction of physical parameters from the hierarchical equation gives better results than the reconstruction from the coupled system, because the errors of the mathematical models have not been taken into account.

Another feature of the inverse problems for the linear models is that accuracy depends on the rate of dispersion of the waves. In almost nondispersive cases, i.e., when $b - \frac{\gamma}{\beta} \approx 0$ in the hierarchical equation or $a_0 - a_1 - \frac{\vartheta}{\alpha} \approx 0$ in the coupled system, results are very bad. For instance, in the case $a_0 = 2.1$, $a_1 = 1$, $\alpha = \vartheta = 10^{-4}$ the relative errors corresponding to $\epsilon = 10^{-3}$ in IPg3 are

$$\left|\frac{a_0^\epsilon - a_0}{a_0}\right| = 577\%, \qquad \left|\frac{a_1^\epsilon - a_1}{a_1}\right| = 0.32\%,$$

$$\left|\frac{\alpha^\epsilon - \alpha}{\alpha}\right| = 27\%, \qquad \left|\frac{\vartheta^\epsilon - \vartheta}{\vartheta}\right| = 820\%.$$

5.5 Proofs of Mathematical Statements

5.5.1 *Proof of Theorem 5.2*

As in the proof of Theorem 5.1 in Sect. 5.1.1, we can make use of the method of vanishing polynomial coefficients. However in the present case we cannot deduce polynomial equations for the single variables $z = k_j^2$ directly from a pair of systems of the form (5.9). The additional fourth order term ω_j^4 makes the immediate algebraic elimination of ω_j impossible. Nevertheless, it is possible to rewrite (5.9) in a form of an algebraic system containing $z = \frac{k_j}{\omega_j}$ and ω_j where the latter could be eliminated from a pair of systems.

Furthermore, in the present case a number k_j may be the value of either $k(\omega_j)$ or $k_2(\omega_j)$. Therefore, we take into consideration a general set of solutions of (4.14). Namely, for any $\omega \in \mathbb{C}$ we define

$$K(\omega) = \{k \in \mathbb{C} : k \text{ solves (4.14) for given } \omega\}.$$

Since (4.14) is a quartic equation, $K(\omega)$ contains maximally 4 elements for any $\omega \in \mathbb{C}$. We split the proof of Theorem 5.2 into lemmas.

Lemma 5.2 *Assume that* $a_0\alpha - a_1\alpha - \vartheta \neq 0$ *and let* $\varkappa_1, \ldots, \varkappa_4$ *be given by* (4.15) *in terms of* $a_0, a_1, \alpha, \vartheta$. *Moreover, let* $\omega_1, \omega_2 \in \mathbb{C}, \omega_1, \omega_2 \neq 0$, *and* $k_j \in K(\omega_j)$, $j = 1, 2$. *If* $\omega_1^2 \neq \omega_2^2$ *then the quotients* $s_j = \frac{k_j}{\omega_j}$ *satisfy* $s_1^2 \neq s_2^2$.

Proof Due to the choice of k_j, the equations

$$\omega_j^4 + \varkappa_1 \omega_j^2 k_j^2 + \varkappa_2 k_j^4 + \varkappa_3 \omega_j^2 + \varkappa_4 k_j^2 = 0 \tag{5.29}$$

hold for $j = 1, 2$. Let $\omega_1^2 \neq \omega_2^2$. Suppose on the contrary that $s_1^2 = s_2^2 =: s^2$. Then, dividing (5.29) by ω_j^4 we have

$$1 + \varkappa_1 s^2 + \varkappa_2 s^4 + \frac{1}{\omega_j^2}\left(\varkappa_3 + \varkappa_4 s^2\right) = 0, \quad j = 1, 2. \tag{5.30}$$

Subtracting these equations for $j = 1$ and $j = 2$ and observing that $\omega_1^2 \neq \omega_2^2$ we get

$$\varkappa_3 + \varkappa_4 s^2 = 0. \tag{5.31}$$

This together with (5.30) gives the equation

$$1 + \varkappa_1 s^2 + \varkappa_2 s^4 = 0. \tag{5.32}$$

Expressing s^2 from (5.31) and substituting into (5.32) we get

$$1 - \varkappa_1 \frac{\varkappa_3}{\varkappa_4} + \varkappa_2 \left(\frac{\varkappa_3}{\varkappa_4}\right)^2 = 0.$$

Using here the formulas (4.15) for $\varkappa_1, \ldots, \varkappa_4$ and simplifying we obtain

$$\frac{\vartheta}{(a_0\alpha - \vartheta)^2}(a_0\alpha - a_1\alpha - \vartheta) = 0.$$

But this relation cannot hold, because $\vartheta > 0$ and $a_0\alpha - a_1\alpha - \vartheta \neq 0$. Therefore, the supposition $s_1^2 = s_2^2$ was not right. We have $s_1^2 \neq s_2^2$ and the lemma is proved. □

We shall prove Theorem 5.2 in the following more general form.

Lemma 5.3 *Assume that* $a_0\alpha - a_1\alpha - \vartheta \neq 0$ *and let* $\omega_j \in \mathbb{C}$, $\omega_j \neq 0$, $j = 1, \ldots, 4$, *be such that* ω_j^2, $j = 1, \ldots, 4$, *are different. Moreover, let us choose some* $k_j \in K(\omega_j)$, $j = 1, \ldots, 4$. *Then the solution of* (5.9) *with the data* (ω_j, k_j), $j = 1, \ldots, 4$, *is unique.*

Proof To prove this assertion, we make use of the method of vanishing polynomial coefficients, again. Suppose that the system (5.9) has two solutions $\varkappa_1, \ldots, \varkappa_4$ and $\widetilde{\varkappa}_1, \ldots, \widetilde{\varkappa}_4$. We write this system up for these solutions and divide by ω_j^4 to get the following equations containing the quotients $s_j = \frac{k_j}{\omega_j}$:

$$1 + \varkappa_1 s_j^2 + \varkappa_2 s_j^4 + \frac{1}{\omega_j^2}\left(\varkappa_3 + \varkappa_4 s_j^2\right) = 0, \quad j = 1, \ldots, 4,$$

$$1 + \widetilde{\varkappa}_1 s_j^2 + \widetilde{\varkappa}_2 s_j^4 + \frac{1}{\omega_j^2}(\widetilde{\varkappa}_3 + \widetilde{\varkappa}_4 s_j^2) = 0, \quad j = 1, \dots, 4.$$

Let us eliminate ω_j from these relations. To this end we multiply the first equations by $\widetilde{\varkappa}_3 + \widetilde{\varkappa}_4 s_j^2$ and the second equations by $\varkappa_3 + \varkappa_4 s_j^2$ and subtract. Then we reach the following expressions:

$$(\varkappa_4 \widetilde{\varkappa}_2 - \widetilde{\varkappa}_4 \varkappa_2) s_j^6 + (\varkappa_3 \widetilde{\varkappa}_2 - \widetilde{\varkappa}_3 \varkappa_2 + \varkappa_4 \widetilde{\varkappa}_1 - \widetilde{\varkappa}_4 \varkappa_1) s_j^4$$
$$+ (\varkappa_4 - \widetilde{\varkappa}_4 + \varkappa_3 \widetilde{\varkappa}_1 - \widetilde{\varkappa}_3 \varkappa_1) s_j^2 + \varkappa_3 - \widetilde{\varkappa}_3 = 0, \quad j = 1, \dots, 4. \tag{5.33}$$

These relations show that $z = s_j^2$, $j = 1, \dots, 4$, are the roots of the following cubic function:

$$f(z) = (\varkappa_4 \widetilde{\varkappa}_2 - \widetilde{\varkappa}_4 \varkappa_2) z^3 + (\varkappa_3 \widetilde{\varkappa}_2 - \widetilde{\varkappa}_3 \varkappa_2 + \varkappa_4 \widetilde{\varkappa}_1 - \widetilde{\varkappa}_4 \varkappa_1) z^2$$
$$+ (\varkappa_4 - \widetilde{\varkappa}_4 + \varkappa_3 \widetilde{\varkappa}_1 - \widetilde{\varkappa}_3 \varkappa_1) z + \varkappa_3 - \widetilde{\varkappa}_3. \tag{5.34}$$

Since ω_j^2, $j = 1, \dots, 4$, are different, by Lemma 5.2 the quantities s_j^2, $j = 1, \dots, 4$, are also different. Consequently, the cubic function (5.34) has four different roots. Thus, it is trivial. Setting the coefficients of (5.34) equal to zero, after some transformations we arrive at the following 4×4 system for the vector $(\widetilde{\varkappa}_1 - \varkappa_1, \widetilde{\varkappa}_2 - \varkappa_2, \widetilde{\varkappa}_3 - \varkappa_3, \widetilde{\varkappa}_4 - \varkappa_4)$:

$$\begin{aligned}
&& \widetilde{\varkappa}_3 - \varkappa_3 && = 0 \\
\varkappa_3(\widetilde{\varkappa}_1 - \varkappa_1) - && \varkappa_1(\widetilde{\varkappa}_3 - \varkappa_3) - && (\widetilde{\varkappa}_4 - \varkappa_4) = 0 \\
\varkappa_4(\widetilde{\varkappa}_1 - \varkappa_1) + \varkappa_3(\widetilde{\varkappa}_2 - \varkappa_2) - && \varkappa_2(\widetilde{\varkappa}_3 - \varkappa_3) - && \varkappa_1(\widetilde{\varkappa}_4 - \varkappa_4) = 0 \\
&& \varkappa_4(\widetilde{\varkappa}_2 - \varkappa_2) - && \varkappa_2(\widetilde{\varkappa}_4 - \varkappa_4) = 0.
\end{aligned}$$

For the determinant of this system we have

$$-\varkappa_2 \varkappa_3^2 - \varkappa_4^2 + \varkappa_1 \varkappa_3 \varkappa_4 = \frac{(a_0\alpha - a_1\alpha - \vartheta)\vartheta}{\delta^2} \neq 0,$$

because $\vartheta > 0$ and $a_0\alpha - a_1\alpha - \vartheta \neq 0$. This implies that the system under consideration has only the trivial solution. Hence, $\widetilde{\varkappa}_1 = \varkappa_1$, $\widetilde{\varkappa}_2 = \varkappa_2$, $\widetilde{\varkappa}_3 = \varkappa_3$, $\widetilde{\varkappa}_4 = \varkappa_4$. The lemma is proved. □

Theorem 5.2 follows from Lemma 5.3 because wavenumbers k_j contained in the data of IPh2 belong to $K(\omega_j)$ for any $j = 1, \dots, 4$.

5.5.2 *Proofs of Sect. 5.2*

Proof of Theorem 5.5 The assertion (ii) immediately follows from Corollary 5.1 and the formula $\frac{1}{c_g} = k'(\omega) = \frac{1}{\sqrt{b}}$ that is valid in the nondispersive case $b\beta - \gamma = 0$ (see Lemma 4.1). Therefore, let us study in detail the dispersive case.

Firstly, we prove the uniqueness for IPg2. Suppose that IPg2 has two solutions: b, β, γ and $\widetilde{b}, \widetilde{\beta}, \widetilde{\gamma}$. As in the proof of Theorem 5.1, from the first two equations of (5.19) we deduce (5.3) for $j = 1, 2$. This means that $z = k_j^2$, $j = 1, 2$, are the roots of the quadratic function $\mathcal{P}_2(z)$ given by (5.4). Further, the third equation of (5.19) in the cases of these two solutions can be written

$$\omega_1(\delta\beta k_1^2 + 1) + (\delta\beta\omega_1^2 - 2\delta\gamma k_1^2 - b)k_1 k_1' = 0, \tag{5.35}$$

$$\omega_1(\delta\widetilde{\beta} k_1^2 + 1) + (\delta\widetilde{\beta}\omega_1^2 - 2\delta\widetilde{\gamma} k_1^2 - \widetilde{b})k_1 k_1' = 0. \tag{5.36}$$

To eliminate k_1', we multiply (5.35) by $\delta\widetilde{\beta}\omega_1^2 - 2\delta\widetilde{\gamma}k_1^2 - \widetilde{b}$, (5.36) by $\delta\beta\omega_1^2 - 2\delta\gamma k_1^2 - b$, subtract and divide by $\omega_1 \neq 0$:

$$(\delta\beta k_1^2 + 1)(\delta\widetilde{\beta}\omega_1^2 - 2\delta\widetilde{\gamma}k_1^2 - \widetilde{b}) - (\delta\widetilde{\beta}k_1^2 + 1)(\delta\beta\omega_1^2 - 2\delta\gamma k_1^2 - b) = 0. \tag{5.37}$$

The next step is the elimination ω_1 from this equation. To this end, we use the first equation in (5.19) in the cases of both solutions:

$$\omega_1^2(\delta\beta k_1^2 + 1) = \delta\gamma k_1^4 + b k_1^2,$$
$$\omega_1^2(\delta\widetilde{\beta} k_1^2 + 1) = \delta\widetilde{\gamma} k_1^4 + \widetilde{b} k_1^2.$$

Applying these relations to ω_1-dependent terms in (5.37) we deduce that

$$\delta\widetilde{\beta}(\delta\gamma k_1^4 + b k_1^2) - (\delta\beta k_1^2 + 1)(2\delta\widetilde{\gamma}k_1^2 + \widetilde{b})$$
$$- \delta\beta(\delta\widetilde{\gamma}k_1^4 + \widetilde{b}k_1^2) + (\delta\widetilde{\beta}k_1^2 + 1)(2\delta\gamma k_1^2 + b) = 0.$$

The latter relation can be rewritten as follows:

$$3\delta^2(\widetilde{\gamma}\beta - \gamma\widetilde{\beta})k_1^4 + 2\delta(\widetilde{\gamma} - \gamma + \widetilde{b}\beta - b\widetilde{\beta})k_1^2 + \widetilde{b} - b = 0. \tag{5.38}$$

Now we subtract from (5.38) the equation (5.3) for $j = 1$ and divide by $k_1^2 \neq 0$. We obtain the following equation:

$$2\delta^2(\widetilde{\gamma}\beta - \gamma\widetilde{\beta})k_1^2 + \delta(\widetilde{\gamma} - \gamma + \widetilde{b}\beta - b\widetilde{\beta}) = 0.$$

This shows that $\mathcal{P}_2'(k_1^2) = 0$. Hence, the number $z = k_1^2$ is a double root of the polynomial $\mathcal{P}_2(z)$. Since $k_1^2 \neq k_2^2$ (this follows from the strict monotonicity of $k(\omega)$ and the inequality $\omega_1^2 \neq \omega_2^2$) we see that the quadratic polynomial $\mathcal{P}_2$ has two different roots k_1^2 and k_2^2, where k_1^2 has the multiplicity 2. This is possible only in case $\mathcal{P}_2$ is the trivial polynomial. Setting the coefficients of $\mathcal{P}_2$ equal to zero, we prove the equalities $\widetilde{b} = b$, $\widetilde{\beta} = \beta$ and $\widetilde{\gamma} = \gamma$ as in the proof of Theorem 5.1. This completes the proof of the uniqueness for IPg2.

The uniqueness for IPg1 can be proved by the same method, i.e., showing that k_0 is a triple root of $\mathcal{P}_2$. However, this is somewhat complicated and involves long computations, because it is necessary to eliminate k_0', k_0'' and ω_0 from related equations. It is easier to use the explicit formula (4.6) for $\omega(k)$ for this purpose, because we have to apply it at a single argument k_0. From (4.6) we have

$$\frac{b+\delta\gamma k^2}{1+\delta\beta k^2}=\left[\frac{\omega(k)}{k}\right]^2. \tag{5.39}$$

By differentiation we deduce that

$$\frac{\gamma-b\beta}{(1+\delta\beta k^2)^2}=\frac{1}{2\delta k}\left\{\left[\frac{\omega(k)}{k}\right]^2\right\}'. \tag{5.40}$$

Differentiating once again we obtain

$$\frac{\beta(\gamma-b\beta)}{(1+\delta\beta k^2)^3}=-\frac{1}{4\delta k}\left[\frac{1}{2\delta k}\left\{\left[\frac{\omega(k)}{k}\right]^2\right\}'\right]'. \tag{5.41}$$

Setting $k=k_0=\frac{\omega_0}{c_{ph}}$, the right-hand sides of (5.39)–(5.41) can be evaluated in terms of the data of IPg1. More precisely, since $\omega(k_0)=\omega_0$, $\omega'(k_0)=c_g$ and $\omega''(k_0)=-k''(\omega_0)[\omega'(k_0)]^3=-2dc_g^3$, we obtain the following system:

$$\frac{b+\delta\gamma k_0^2}{1+\delta\beta k_0^2}=c_{ph}^2, \tag{5.42}$$

$$\frac{\gamma-b\beta}{(1+\delta\beta k_0^2)^2}=r_1 \tag{5.43}$$

$$\frac{\beta(\gamma-b\beta)}{(1+\delta\beta k_0^2)^3}=r_2 \tag{5.44}$$

where

$$r_1=\frac{c_{ph}}{\delta k_0^2}(c_g-c_{ph}),\qquad r_2=-\frac{1}{4\delta^2k_0^4}\left[(c_g-4c_{ph})(c_g-c_{ph})-2dc_g^3c_{ph}k_0\right].$$

Dividing (5.43) by (5.44) we evaluate $\beta=[\frac{r_1}{r_2}-\delta k_0^2]^{-1}$. Once β is known, from (5.42) and (5.43) a 2×2 linear system for b and γ can be constructed:

$$b+\delta k_0^2\gamma=c_{ph}^2\left(1+\delta\beta k_0^2\right),$$

$$-\beta b+\gamma=r_1\left(1+\delta\beta k_0^2\right)^2.$$

The determinant of this system is $1+\delta\beta k_0^2$ and it differs from zero because $\delta,\beta>0$ (see (3.37)). Thus, the solution the linear system is unique. Summing up, the solution b,β,γ of IPg1 is unique. The theorem is proved. □

Proof of Theorem 5.6 Suppose that (5.21) has two solutions $\varkappa_1, \dots, \varkappa_4$ and $\widetilde{\varkappa}_1, \dots, \widetilde{\varkappa}_4$. This means that the following equalities hold:

$$k_j^2\omega_j^2\varkappa_1 + k_j^4\varkappa_2 + \omega_j^2\varkappa_3 + k_j^2\varkappa_4 = -\omega_j^4, \quad j = 1, 2,$$
$$k_j^2\omega_j^2\widetilde{\varkappa}_1 + k_j^4\widetilde{\varkappa}_2 + \omega_j^2\widetilde{\varkappa}_3 + k_j^2\widetilde{\varkappa}_4 = -\omega_j^4, \quad j = 1, 2,$$
$$\left(\omega_j k_j^2 + \omega_j^2 k_j k_j'\right)\varkappa_1 + 2k_j^3 k_j'\varkappa_2 + \omega_j\varkappa_3 + k_j k_j'\varkappa_4 = -2\omega_j^3, \quad j = 1, 2,$$
$$\left(\omega_j k_j^2 + \omega_j^2 k_j k_j'\right)\widetilde{\varkappa}_1 + 2k_j^3 k_j'\widetilde{\varkappa}_2 + \omega_j\widetilde{\varkappa}_3 + k_j k_j'\widetilde{\varkappa}_4 = -2\omega_j^3, \quad j = 1, 2.$$

Dividing the first two equalities by ω_j^4 and the last two equalities by ω_j^3 and denoting $s_j = \frac{k_j}{\omega_j}$ we obtain

$$\begin{aligned}
&1 + \varkappa_1 s_j^2 + \varkappa_2 s_j^4 + \frac{1}{\omega_j^2}\left(\varkappa_3 + \varkappa_4 s_j^2\right) = 0,\\
&1 + \widetilde{\varkappa}_1 s_j^2 + \widetilde{\varkappa}_2 s_j^4 + \frac{1}{\omega_j^2}\left(\widetilde{\varkappa}_3 + \widetilde{\varkappa}_4 s_j^2\right) = 0,\\
&2 + \varkappa_1 s_j^2 + \frac{\varkappa_3}{\omega_j^2} + \left(\varkappa_1 + 2\varkappa_2 s_j^2 + \frac{\varkappa_4}{\omega_j^2}\right)s_j k_j' = 0,\\
&2 + \widetilde{\varkappa}_1 s_j^2 + \frac{\widetilde{\varkappa}_3}{\omega_j^2} + \left(\widetilde{\varkappa}_1 + 2\widetilde{\varkappa}_2 s_j^2 + \frac{\widetilde{\varkappa}_4}{\omega_j^2}\right)s_j k_j' = 0,
\end{aligned} \tag{5.45}$$

where $j = 1, 2$. As in the proof of Lemma 5.3, the elimination of ω_j from the first two equations in (5.45) leads to expression (5.33). This shows that s_1^2 and s_2^2 are roots of the cubic function $f(z)$ defined by (5.34).

There is another possibility for eliminating k_j' and ω_j from (5.45), too. Namely, let us multiply the fourth equation by $\varkappa_1 + 2\varkappa_2 s_j^2 + \frac{\varkappa_4}{\omega_j^2}$, the third equation by $\widetilde{\varkappa}_1 + 2\widetilde{\varkappa}_2 s_j^2 + \frac{\widetilde{\varkappa}_4}{\omega_j^2}$ and subtract to get rid of k_j':

$$\begin{aligned}
&\left(2 + \varkappa_1 s_j^2 + \frac{\varkappa_3}{\omega_j^2}\right)\left(\widetilde{\varkappa}_1 + 2\widetilde{\varkappa}_2 s_j^2 + \frac{\widetilde{\varkappa}_4}{\omega_j^2}\right)\\
&\quad - \left(2 + \widetilde{\varkappa}_1 s_j^2 + \frac{\widetilde{\varkappa}_3}{\omega_j^2}\right)\left(\varkappa_1 + 2\varkappa_2 s_j^2 + \frac{\varkappa_4}{\omega_j^2}\right) = 0, \quad j = 1, 2.
\end{aligned} \tag{5.46}$$

Further, we multiply the second equation in (5.45) by $2\varkappa_1 + 4\varkappa_1 s_j^2 + \frac{\varkappa_4}{\omega_j^2}$, the first equation by $2\widetilde{\varkappa}_1 + 4\widetilde{\varkappa}_1 s_j^2 + \frac{\widetilde{\varkappa}_4}{\omega_j^2}$ and subtract again. The result is

$$\left[1+\varkappa_1 s_j^2+\varkappa_2 s_j^4+\frac{1}{\omega_j^2}(\varkappa_3+\varkappa_4 s_j^2)\right]\left(2\widetilde{\varkappa}_1+4\widetilde{\varkappa}_1 s_j^2+\frac{\widetilde{\varkappa}_4}{\omega_j^2}\right)$$
$$-\left[1+\widetilde{\varkappa}_1 s_j^2+\widetilde{\varkappa}_2 s_j^4+\frac{1}{\omega_j^2}(\widetilde{\varkappa}_3+\widetilde{\varkappa}_4 s_j^2)\right]\left(2\varkappa_1+4\varkappa_1 s_j^2+\frac{\varkappa_4}{\omega_j^2}\right)=0,$$
$$j=1,2. \tag{5.47}$$

Finally, subtracting (5.46) from (5.47), only terms with the factor $\frac{1}{\omega_j^2}$ remain:

$$\frac{1}{\omega_j^2}\big[(\varkappa_3+\varkappa_4 s_j^2)(2\widetilde{\varkappa}_1+4\widetilde{\varkappa}_2 s_j^2)+\widetilde{\varkappa}_4(1+\varkappa_1 s_j^2+\varkappa_2 s_j^4)$$
$$-(\widetilde{\varkappa}_3+\widetilde{\varkappa}_4 s_j^2)(2\varkappa_1+4\varkappa_2 s_j^2)-\varkappa_4(1+\widetilde{\varkappa}_1 s_j^2+\widetilde{\varkappa}_2 s_j^4)\big]$$
$$-\frac{1}{\omega_j^2}\big[\varkappa_3(\widetilde{\varkappa}_1+2\widetilde{\varkappa}_2 s_j^2)+\widetilde{\varkappa}_4(2+\varkappa_1 s_j^2)$$
$$-\widetilde{\varkappa}_3(\varkappa_1+2\varkappa_2 s_j^2)-\varkappa_4(2+\widetilde{\varkappa}_1 s_j^2)\big]=0, \quad j=1,2.$$

Multiplying by $\omega_j^2 \neq 0$ and simplifying we obtain

$$3(\varkappa_4\widetilde{\varkappa}_2-\widetilde{\varkappa}_4\varkappa_2)s_j^4+2(\varkappa_3\widetilde{\varkappa}_2-\widetilde{\varkappa}_3\varkappa_2+\varkappa_4\widetilde{\varkappa}_1-\widetilde{\varkappa}_4\varkappa_1)s_j^2$$
$$+\varkappa_4-\widetilde{\varkappa}_4+\varkappa_3\widetilde{\varkappa}_1-\widetilde{\varkappa}_3\varkappa_1=0, \quad j=1,2.$$

From these relations we have $f'(s_j^2)=0$ for $j=1,2$. This means that s_j^2, $j=1,2$, are double roots of the cubic function $f(\sigma)$. Since ω_j^2, $j=1,2$, are different, by Lemma 5.2 the quantities s_j^2, $j=1,2$, are also different. Therefore, the cubic function $f(\sigma)$ has two different double roots and hence it is trivial. The rest of the proof is identical to that of Lemma 5.3. □

Chapter 6
Solitary Waves in Nonlinear Models

6.1 Solitary Waves

In many physical problems there is a long list of phenomena which influence the possible output and which should be taken into account in adequate mathematical models. Here we focus our attention on the competing nonlinearity and dispersion in wave motion. It is well known that if these effects are balanced then solitary waves may emerge. Discovered first by John Scott Russell [62] in a "natural experiment" with waves in a narrow canal, the solitary waves were theoretically found a half a century later as steady state solutions to shallow water equations [42]. More than the next half a century later, the quest for the energy equipartition in lattices (the Fermi–Pasta–Ulam problem) stimulated more studies which involved also the continuum limit to lattice equations. This was the way that Zabusky and Kruskal [72] "reinvented" the Korteweg–de Vries (KdV) equation and demonstrated the emergence of steady solitary waves from a harmonic input. They also coined the term "soliton". Nowadays solitary waves and solitons form a paradigm in many branches of physics (see, for example [8]) including wave propagation in solids. The celebrated KdV equation reads

$$u_t + k\left(u^2\right)_x + d u_{xxx} = 0 \tag{6.1}$$

where u is a field variable and x, t are independent variables reflecting the moving space coordinate and time, respectively; k and d are constants. This equation admits a soliton-type sech^2 solution which describes a wave moving either to the right or to the left depending on the choice of the moving coordinate x. Solitons emerge due to the balance of quadratic nonlinearity and cubic dispersion—a classical case nowadays. The derivation of (6.1) and its modifications are described in detail by Taniuti and Nishihara [66] and Engelbrecht [11, 12].

It is not only the KdV equation which admits the soliton-type solutions—see for example [8]. However, here we are interested only in waves in solids and focus our attention on models like (6.1) and its counterparts which in our case are of a different form. We need also some working definitions [12]:

J. Janno, J. Engelbrecht, *Microstructured Materials: Inverse Problems*, Springer Monographs in Mathematics,

DOI 10.1007/978-3-642-21584-1_6,

Definition 1 Solitary pulse waves are progressive steady waves the profiles of which describe a smooth transition from an equilibrium state to the same equilibrium state.

Definition 2 Solitons are solitary waves which conserve their profiles and velocities during the collisions with other solitons.

It is proved that the solitary waves described by the KdV equation are solitons. We leave aside their remarkable properties and note here only that the KdV equation conserves the energy. The situation, however, can be much more complicated due to the more complicated nonlinear and dispersive effects, and in these cases the solitary waves, either generated or emerging, do not satisfy Definition 2, although they satisfy Definition 1. It means also that their profiles are not of the sech^2 type.

Another important aspect is that the overwhelming majority of studies on solitary waves and/or solitons are based on one-wave equations like (6.1). At the same time, even the classical one-dimensional wave equation describes two waves—one propagating to the right, another—to the left. There is a considerable interest to analyse the existence and emergence of solitary waves as solutions to two-wave equations. In this book, the corresponding mathematical models were derived in Chap. 3. Equation (3.36) reads:

$$v_{tt} = b v_{xx} + \cdots \tag{6.2}$$

which describes indeed two waves like a classical wave equation but reflects more complicated properties—complicated dispersion and nonlinearities at the macro- and micro-levels. Such models are analysed numerically by Salupere et al. [59] but we need also an analytical treatment in order to solve the inverse problems.

6.2 Solitary Wave Solutions of Hierarchical Equation

We start by studying the hierarchical equation in the nonlinear case that was derived in Sect. 3.2:

$$v_{tt} = b v_{xx} + \frac{\mu}{2}\left(v^2\right)_{xx} + \delta(\beta v_{tt} - \gamma v_{xx})_{xx} + \delta^{3/2}\frac{\lambda}{2}\left(v_x^2\right)_{xxx}. \tag{6.3}$$

Travelling wave solutions of (6.3) are of the form

$$v(x,t) = w(x - ct), \tag{6.4}$$

where c is a free parameter (velocity of the wave) and $w = w(\xi)$ is a solution of the equation

$$\left(c^2 - b\right)w'' - \frac{\mu}{2}\left(w^2\right)'' - \delta\left(\beta c^2 - \gamma\right)w^{IV} - \delta^{3/2}\frac{\lambda}{2}\left[\left(w'\right)^2\right]''' = 0. \tag{6.5}$$

We treat (6.5) in the classical sense. This means that we require the solution to be four times continuously differentiable. Moreover, we search for *solitary wave* solutions, i.e., solutions, which are nontrivial and vanish at infinity. Due to these requirements, we define the following set of admissible solutions for (6.5):

$$\mathcal{W}_4 = \{ w \in C^4(\mathbb{R}) : w \not\equiv 0 \text{ and } w^{(j)}(\xi) \to 0 \text{ as } |\xi| \to \infty,\ j = 0, 1, 2, 3 \}. \tag{6.6}$$

Here $C^k(D)$ denotes the space of k times continuously differentiable functions on the set D.

The aim of this section is to give a mathematically rigorous explanation of the existence and properties of solitary waves in materials characterised by (6.5).

6.2.1 Reduction to Equation of First Kind. Canonical Description

Let us start by integrating twice (6.5). This leads to the equivalent equation

$$(c^2 - b)w - \frac{\mu}{2}w^2 - \delta(\beta c^2 - \gamma)w'' - \delta^{3/2}\frac{\lambda}{2}[(w')^2]' = C_1\xi + C_2 \tag{6.7}$$

with arbitrary constants C_1 and C_2. Due to the vanishing conditions in the definition of $\mathcal{W}_4$, we have $C_1 = C_2 = 0$. Therefore, (6.5) is in the space $\mathcal{W}_4$ equivalent to the equation of the second order

$$(c^2 - b)w - \frac{\mu}{2}w^2 - [\delta(\beta c^2 - \gamma) + \delta^{3/2}\lambda w']w'' = 0. \tag{6.8}$$

Lemma 6.1 *Let* (6.5) *have a solution in the space* $\mathcal{W}_4$. *Then* $\beta c^2 - \gamma \neq 0$. *Moreover,* $w'(\xi) \neq 0$ *a.e. and* $\delta(\beta c^2 - \gamma) + \delta^{3/2}\lambda w'(\xi) \neq 0$ *a.e.*

Here and in the sequel *a.e.* is the abbreviation of the phrase "*almost everywhere*". We say that a certain relation holds *a.e.* if it may fail maximally on a countable set of numbers.

The proof of Lemma 6.1 is shifted to Sect. 6.4.

Due to the assertion about $w'(\xi)$ in Lemma 6.1, (6.8) is equivalent to the following equation obtained by multiplication by w':

$$w''[\delta(\beta c^2 - \gamma)w' + \delta^{3/2}\lambda(w')^2] = \left[(c^2 - b)w - \frac{\mu}{2}w^2\right]w'.$$

Let us integrate this equation once again taking the behaviour $w(\xi), w'(\xi) \to 0$ as $|\xi| \to \infty$ into account. This results in the following equation of the first order that is equivalent to (6.5) in the space $\mathcal{W}_4$:

$$\frac{\delta(\beta c^2 - \gamma)}{2}(w')^2 + \frac{\delta^{3/2}\lambda}{3}(w')^3 = \frac{c^2 - b}{2}w^2 - \frac{\mu}{6}w^3.$$

In view of $\beta c^2 - \gamma \neq 0$, we transform it to the form

$$(w')^2 + \frac{2\delta^{1/2}\lambda}{3(\beta c^2 - \gamma)}(w')^3 = \frac{c^2 - b}{\delta(\beta c^2 - \gamma)}w^2 - \frac{\mu}{3\delta(\beta c^2 - \gamma)}w^3. \tag{6.9}$$

We can deduce further necessary conditions for the coefficients (proof is in Sect. 6.4).

Lemma 6.2 *Let* (6.5) *have a solution in the space* $\mathcal{W}_4$. *Then* $c^2 - b \neq 0$, $\mu \neq 0$ *and* $\frac{c^2-b}{\delta(\beta c^2-\gamma)} > 0$.

Let us define the following three parameters which have certain physical or geometrical meanings:

$$\kappa := \sqrt{\frac{c^2 - b}{\delta(\beta c^2 - \gamma)}}, \qquad A := \frac{3(c^2 - b)}{\mu}, \qquad \Theta := -2\left[\frac{c^2 - b}{\beta c^2 - \gamma}\right]^{3/2}\frac{\lambda}{\mu}. \tag{6.10}$$

With these parameters equation (6.9) has the form

$$(w')^2 - \frac{\Theta}{\kappa A}(w')^3 = \kappa^2 w^2\left(1 - \frac{w}{A}\right). \tag{6.11}$$

Equation (6.11) admits the form of the autonomous system of the first order, too:

$$w' = W, \qquad W' = \frac{(c^2 - b)w - \frac{\mu}{2}w^2}{\delta(\beta c^2 - \gamma) + \delta^{3/2}\lambda W}. \tag{6.12}$$

Here the denominator $\delta(\beta c^2 - \gamma) + \delta^{3/2}\lambda W = \delta(\beta c^2 - \gamma) + \delta^{3/2}\lambda w'$ is not identically zero in view of Lemma 6.1.

The solution of (6.11) depends upon the parameters κ, A and Θ. Here κ is the exponential decay rate of the solution. This can be seen comparing (6.11) with the definition of κ. We obtain

$$(w')^2 \sim \kappa^2 w^2 \quad \text{as } |\xi| \to \infty, \tag{6.13}$$

which implies

$$\ln|w(\xi)| \sim -\kappa|\xi| \quad \text{as } |\xi| \to \infty. \tag{6.14}$$

The inverse $1/\kappa$ is usually called the width of the wave because it is proportional to the width of the observable support of the wave. Further we will see that A is the amplitude of the wave. The parameter Θ is related to the asymmetry of the wave. The size of Θ which depends on the ratio $\frac{\lambda}{\mu}$ of coefficients of nonlinear terms of the wave equation, is important from the point of view of the existence of the solitary wave. We will study this issue closely in Sect. 6.2.2 making use of the geometry of trajectories of the equation in the phase plane. The physical background will be given in Sect. 6.2.3.

To simplify the study of (6.11), we define new variables

$$y = \frac{1}{A} w, \qquad \zeta = \kappa \xi. \tag{6.15}$$

Then (6.11) is reduced to the following *canonical* form:

$$\left(y'\right)^2 - \Theta \left(y'\right)^3 = y^2 - y^3. \tag{6.16}$$

Here $y' = \frac{dy}{d\zeta}$. For further study, we solve it with respect to y':

$$y' = Q(y). \tag{6.17}$$

Then the inverse of the solution $y(\zeta)$ has the form

$$\zeta = \int \frac{dy}{Q(y)}. \tag{6.18}$$

Unfortunately, the analytical solution of (6.17) is very complicated because it involves an integration of a cubic function in terms of another cubic function. To the authors' knowledge, this integral cannot be evaluated within known functions in the general case. Nevertheless, when the nonlinearity in the microscale is absent, i.e., $\lambda = \Theta = 0$, the integration is simple. Then we reach a symmetric bell-shaped solitary wave in the explicit form

$$y(\zeta) = \cosh^{-2}\left(\frac{\zeta}{2}\right) \quad \Longrightarrow \quad w(\xi) = A \cosh^{-2}\left(\frac{\kappa \xi}{2}\right). \tag{6.19}$$

In canonical case the autonomous system (6.12) reads

$$y' = z, \qquad z' = \frac{y(2-3y)}{2 - 3\Theta z}. \tag{6.20}$$

The systems (6.12) and (6.20) can be used in numerical solution of the problem.

6.2.2 Existence and Basic Properties of Canonical Waves

We are going to study the canonical equation (6.16). Let us first consider the case $\Theta \geq 0$. Observing that the solitary wave solution of (6.16) satisfies the conditions $y, y' \to 0$ as $|\zeta| \to \infty$, we see that the trajectory (phase curve) T of (6.20), corresponding to this solution, satisfies the following condition:

$$T \text{ is a closed curve containing the point } O = (0,0). \tag{6.21}$$

To locate such a trajectory in the phase plane, we denote by $\boldsymbol{\phi} = (\phi_1, \phi_2)$ the right-hand side of system (6.20), i.e., $\phi_1 = z$, $\phi_2 = \frac{y(2-3y)}{2-3\Theta z}$ and investigate the sign of components of the right-hand side of (6.20).

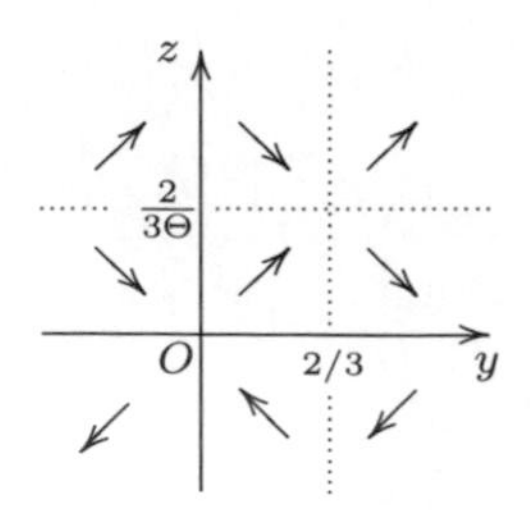

Fig. 6.1 Phase portrait of (6.20)

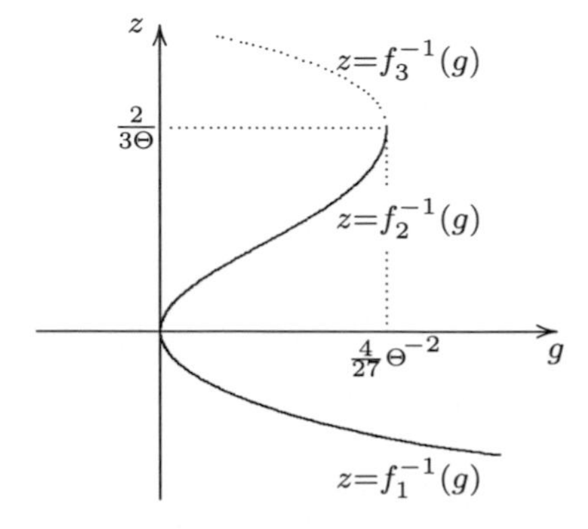

Fig. 6.2 Function $f^{-1}(g)$

We call a *critical line* of an autonomous system a line in the phase plane such that, on crossing this line, at least one of the components of the right-hand side vector of the system changes sign.

Now we observe that the critical lines of (6.20) are the zero lines $y = 0$, $y = \frac{2}{3}$, $z = 0$ and the singularity line $z = \frac{2}{3\Theta}$. They divide the phase plane into 9 subregions. The vector $\boldsymbol{\phi}$ preserves its orientation in each of these subdomains. Figure 6.1 shows the corresponding phase portrait. Due to the orientation of $\boldsymbol{\phi}$, a trajectory T with the property (6.21) can be potentially found only in the quarter $y \geq 0$, $z < \frac{2}{3\Theta}$.

Due to (6.16), the equation of T reads $z^2 - \Theta z^3 = y^2 - y^3$. Our plan is to study this equation in the quarter $y \geq 0$, $z < \frac{2}{3\Theta}$. For this purpose we express it as $z = f^{-1}(y^2 - y^3)$ where f^{-1} is the inverse of $f(z) = z^2 - \Theta z^3$. To analyse the behaviour of this function we introduce the intermediate variable g and split the equation $z = f^{-1}(y^2 - y^3)$ into two subsequent relations

$$z = f^{-1}(g), \quad g = y^2 - y^3.$$

The components $f^{-1}(g)$ and $g(y)$ are shown in Figs. 6.2 and 6.3.

The inverse f^{-1} contains three branches f_1^{-1}, f_2^{-1} and f_3^{-1}. The latter branch cannot be related to the solitary wave because it falls beyond the singularity line $z = \frac{2}{3\Theta}$. The other branches f_1^{-1} and f_2^{-1} are defined for nonnegative values of g. This property together with the above inequality $y \geq 0$ restricts the domain of g to $[0, 1]$. The branch f_1^{-1} yields the curve $z = f_1^{-1}(y^2 - y^3)$, which connects the points $(0, 0)$ and $(1, 0)$ and is located in the lower half-plane $z < 0$ (lower parts of the trajectories in Figs. 6.4, 6.5, 6.6).

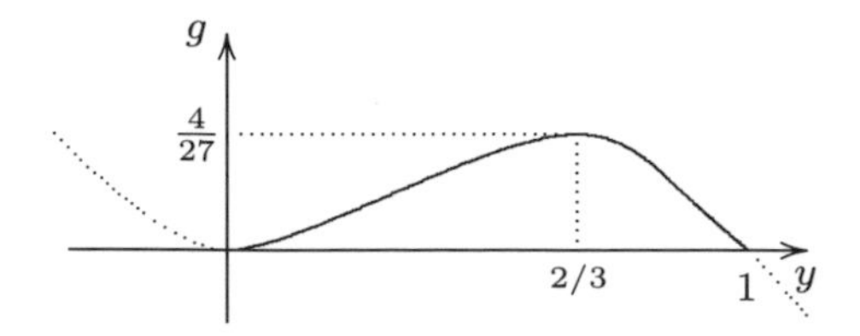

Fig. 6.3 Function $g = y^2 - y^3$

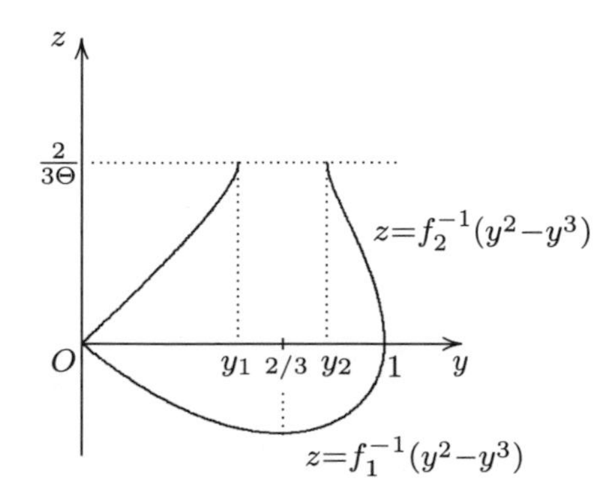

Fig. 6.4 T in case $\Theta > 1$

Concerning the branch f_2^{-1}, three different cases can occur:

(1) $\Theta > 1$. Let us compare the range $[0, \frac{4}{27}]$ of g with the domain $[0, \frac{4}{27\Theta^2}]$ of f_2^{-1}. In view of the inequality $\frac{4}{27\Theta^2} < \frac{4}{27}$, the range of g extends beyond the domain of f_2^{-1}. Moreover, from Fig. 6.3 we see that if we restrict the range of g to the interval $[0, \frac{4}{27\Theta^2}]$, the domain of g is restricted to a certain set of the form $[0, y_1] \cup [y_2, 1]$, where y_1 and y_2 are certain numbers in the intervals $(0, \frac{2}{3})$ and $(\frac{2}{3}, 1)$, respectively. Therefore, the composition $z = f_2^{-1}(g(y)) = f_2^{-1}(y^2 - y^3)$ is defined for any $y \in [0, y_1] \cup [y_2, 1]$ but not for $y \in (y_1, y_2)$. This case is depicted in Fig. 6.4. System (6.20) has no trajectory satisfying the property (6.21). Consequently, the solitary wave does not exist.

(2) $\Theta = 1$. Then the curve $z = f_2^{-1}(y^2 - y^3)$ connects the points $(0, 0)$ and $(1, 0)$ and is located on the upper half-plane $z > 0$ (see Fig. 6.5). The function $z = f_2^{-1}(y^2 - y^3)$ has the maximum point $(\frac{2}{3}, \frac{2}{3})$ on the singularity line $z = \frac{2}{3}$. To study the behaviour of the curve at this point, we note that the equation of the trajectory is $z^2 - z^3 = y^2 - y^3$. This equation admits a particular linear solution $z = y$ passing through $(0, 0)$. This means that the curve $z = f_2^{-1}(y^2 - y^3)$ is the straight line $z = y$ to the left of $y = \frac{2}{3}$. It has a positive slope at the point $(\frac{2}{3}, \frac{2}{3})$. Therefore, the function $z = f_2^{-1}(y^2 - y^3)$ is not smooth at $(\frac{2}{3}, \frac{2}{3})$. This implies that the function $\frac{dz}{dy}$ is discontinuous, which in turn yields that y'' is also discontinuous. A solitary wave solution does not exist in $\mathcal{W}_4$. Nevertheless, the solution exists in a certain generalised sense.

(3) $0 \le \Theta < 1$. Then the relation $\frac{4}{27\Theta^2} > \frac{4}{27}$ holds. The curve $z = f_2^{-1}(y^2 - y^3)$ connects the points $(0, 0)$ and $(1, 0)$ and is located in the band $0 < z < \frac{2}{3\Theta}$. This

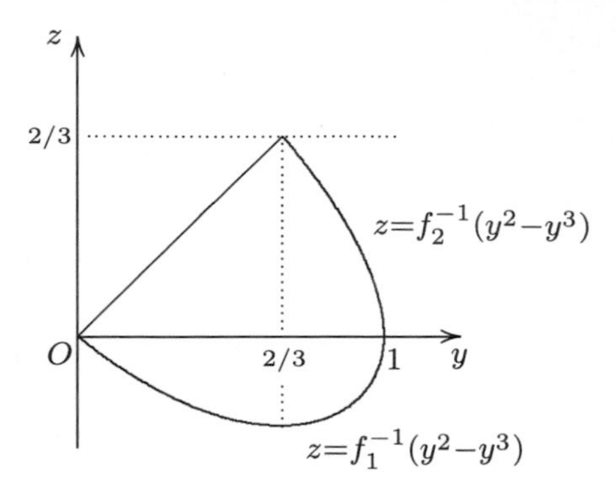

Fig. 6.5 T in case $\Theta = 1$

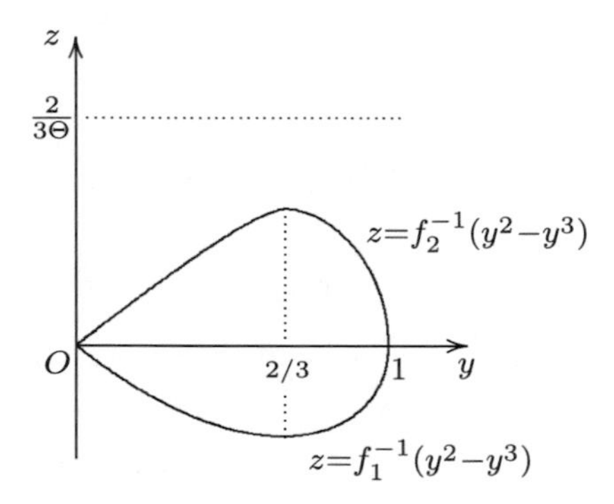

Fig. 6.6 T in case $0 \leq \Theta < 1$

case is shown in Fig. 6.6. The trajectory T is defined as the union of the curves $z = f_1^{-1}(y^2 - y^3)$ and $z = f_2^{-1}(y^2 - y^3)$. It has the property (6.21).

Let us consider the Cauchy problem for the ODE system (6.20) with the conditions $y(0) = 1, z(0) = 0$. By Cauchy theorem, this problem has a solution (y, z). Due the relation $z^2 \sim y^2$ as $y \to 0$ following from the equation $z^2 - \Theta z^3 = y^2 - y^3$, this solution satisfies the conditions $y(\zeta), z(\zeta) = y'(\zeta) \to 0$ as $|\zeta| \to \infty$. Since the right-hand side of (6.20) is infinitely differentiable for $y \geq 0, z < \frac{2}{3\Theta}$, the solution component y is also infinitely differentiable. Therefore, y is an element of $\mathcal{W}_4$. This means that y is the desired solitary wave solution. Other solitary wave solutions can be deduced from $y(\zeta)$ by the argument shift $\zeta \mapsto \zeta + C$, where C is a constant.

Qualitative properties (monotonicity and convexity intervals, etc.) of $y(\zeta)$ can be immediately obtained from the related properties of the trajectory T. More precisely, it holds

$$\left.\begin{aligned} &y \text{ is positive and attains the maximal value } 1;\\ &y \text{ is increasing in case } \zeta < 0 \text{ and decreasing in case} \zeta > 0;\\ &\text{there exist } \zeta_1 < 0 \text{ and } \zeta_2 > 0 \text{ such that } y \text{ is concave}\\ &\text{in case } \zeta < \zeta_1, \zeta > \zeta_2 \text{ and convex in case } \zeta_1 < \zeta < \zeta_2;\\ &y(\zeta_1) = y(\zeta_2) = \frac{2}{3}. \end{aligned}\right\} \quad (6.22)$$

From (6.16) it follows that $y(\zeta)$ solves (6.16) if and only if $y(-\zeta)$ solves (6.16) with Θ replaced by $-\Theta$. Consequently, the solution corresponding to $\Theta < 0$ is the reflection over the line $\zeta = 0$ of the solution corresponding to $\Theta > 0$.

Summing up, we have proved the following theorem.

Theorem 6.1 *Equation* (6.16) *has a solution in* $\mathcal{W}_4$ *if and only if*

$$|\Theta| < 1. \quad (6.23)$$

The set of all solutions in $\mathcal{W}_4$ *has the form* $\{y_C(\xi) = y_0(\xi + C) : C \in \mathbb{R}\}$, *where* $y = y_0 \in \mathcal{W}_4$ *has the properties* (6.22).

From the behaviour of the phase curve we derive an estimate for $y'(\zeta)$ that we will use in further analysis. To this end, let us return to the case $\Theta > 0$. From the definition of the branches f_1^{-1} and f_2^{-1} we obtain $|f_1^{-1}(g)| < |f_2^{-1}(g)|$ for any g. Therefore, the phase curve is located in the band $|z| < \frac{2}{3\Theta}$ (Fig. 6.6). This yields the inequality $|y'(\zeta)| < \frac{2}{3\Theta}$ for any $\zeta \in \mathbb{R}$. In case $\Theta < 0$ we analogously deduce that $|y'(\zeta)| < -\frac{2}{3\Theta}$ for any $\zeta \in \mathbb{R}$. Consequently,

$$\left|y'(\zeta)\right| < \frac{2}{3|\Theta|} \quad \text{for any } \zeta \in \mathbb{R}. \quad (6.24)$$

Two examples of the solitary wave solutions, computed numerically by means of the second order Adams-Bashforth method, are shown in Figs. 6.7 and 6.8.

We remark that the wave is asymmetric in case $\Theta \neq 0$. Let us try to measure this asymmetry. To this end, we note that the solution $y(\zeta)$ is strictly monotone to the left and right of the amplitude point $\zeta = 0$. Consequently, it has two inverses: $\zeta^-(y) < 0$ and $\zeta^+(y) > 0$ that are defined for any $y \in (0, 1)$. Here $|\zeta^-(y)|$ and $|\zeta^+(y)|$ are the front and rear half-lengths of the wave at the fixed level y. Therefore, we can measure the asymmetry of the wave at level $y \in (0, 1)$ by the following ratio of the half-lengths

$$\frac{|\zeta^+(y)|}{|\zeta^-(y)|}. \quad (6.25)$$

It turns out that this ratio is an increasing function of Θ. More precisely, the following statement is valid.

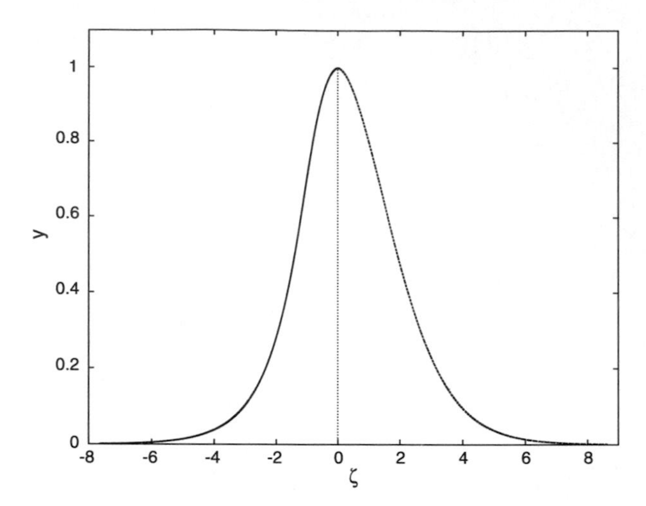

Fig. 6.7 Canonical wave in case $\Theta = 0.9$

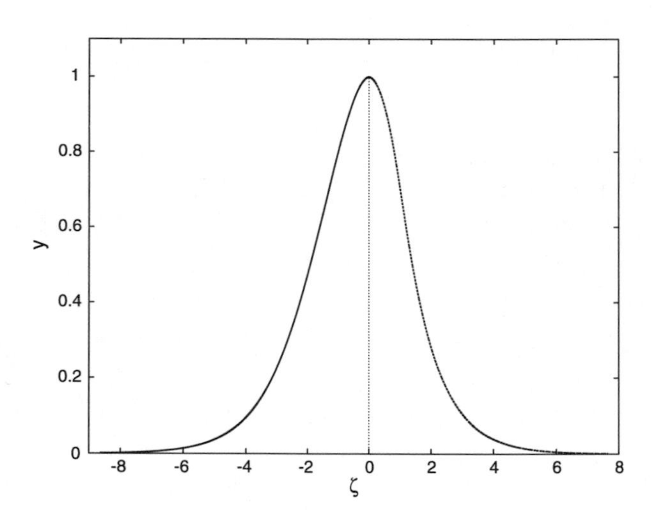

Fig. 6.8 Canonical wave in case $\Theta = -0.9$

Theorem 6.2 *For any* $y \in (0, 1)$ *the relation*

$$\frac{|\zeta^+(y)|}{|\zeta^-(y)|} = F_y(\Theta) \tag{6.26}$$

is valid, where $F_y(\Theta)$ *is an increasing function of* Θ *in the interval* $(-1, 1)$. *Moreover,* $F_y(0) = 1$.

Proof is included in Sect. 6.4.

6.2.3 *Physical and Geometrical Properties of Solitary Waves in General Form*

Let us return to the non-canonical equation (6.11) that contains the parameters A, κ and Θ. In view of the results of the previous section, the equation has a solitary wave solution if and only if $|\Theta| < 1$. Due to (6.15) this solution has the form

$$w(\xi) = A y_\Theta(\kappa\xi), \tag{6.27}$$

where y_Θ is the canonical solitary wave corresponding to Θ.

Evidently, $A = \frac{3(c^2-b)}{\mu}$ is the amplitude of the wave. Depending on the signs of $c^2 - b$ and μ, the amplitude of the wave may be positive or negative. The absolute value of the amplitude is an increasing function of c^2.

Further, the necessary and sufficient solvability condition $|\Theta| < 1$ in terms of the coefficients of the non-canonical equation reads

$$\left(\frac{\beta c^2 - \gamma}{c^2 - b}\right)^3 > \frac{4\lambda^2}{\mu^2}. \tag{6.28}$$

Now we can formulate a theorem concerning the existence and main geometrical properties of w. This immediately follows from the discussions of the previous section, in particular from Theorem 6.1 and the formula (6.24).

Theorem 6.3 *Let $\mu \neq 0$, $\beta c^2 - \gamma \neq 0$ and $c^2 - b \neq 0$. Equation* (6.5) *has a solution in $\mathcal{W}_4$ if and only if the inequality* (6.28) *is satisfied. The set of all solutions in $\mathcal{W}_4$ has the form $\{w_C(\xi) = w_0(\xi + C) : C \in \mathbb{R}\}$, where $w = w_0 \in \mathcal{W}_4$ is an infinitely differentiable function in $\mathbb{R}$, which has the following properties*:

(a) $A^{-1}w(\xi) \in (0, 1)$ *if* $\xi \neq 0$ *and* $w(0) = A$;
(b) $Aw'(\xi) > 0$ *if* $\xi < 0$, $Aw'(\xi) < 0$ *if* $\xi > 0$ *and* $w'(0) = 0$; w' *has exactly two relative extrema occurring at points* $\xi = \xi^- < 0$ *and* $\xi = \xi^+ > 0$ *such that* $w(\xi^-) = w(\xi^+) = \frac{2A}{3}$;
(c) $|w'(\xi)| < \frac{2\kappa|A|}{3|\Theta|}$.

The parameter $\Theta = -2[\frac{c^2-b}{\beta c^2-\gamma}]^{3/2}\frac{\lambda}{\mu}$ is related to the asymmetry. Noting that the sign of Θ equals the sign of $-\mu\lambda$, the following subcases may occur.

(1) In case $A > 0$, $\mu\lambda < 0$ the wave has the shape of the wave in Fig. 6.7.
(2) In case $A > 0$, $\mu\lambda > 0$ the wave has the shape of the wave in Fig. 6.8.
(3) In case of negative A the wave is the reflection over the line $w = 0$ of the wave corresponding to the amplitude $-A$.

The wave function $w(\xi)$ is always strictly monotone to the left and right of the amplitude point $\xi = 0$. Therefore, $w(\xi)$ has two inverses: $\xi^-(w) < 0$ and $\xi^+(w) > 0$ which are defined for any w between 0 and A. Let us choose some relative level $y \in (0, 1)$ and consider the front and rear half-lengths of the wave at this relative level, i.e., the quantities $|\xi^-(yA)|$ and $|\xi^+(yA)|$. The asymmetry of the wave at the relative level $y \in (0,1)$ is equal to the ratio

$$\frac{|\xi^+(yA)|}{|\xi^-(yA)|}. \tag{6.29}$$

Observing that the relation $\xi^{\pm}(yA) = \frac{1}{\kappa}\zeta^{\pm}(y)$ is valid between the inverses of non-canonical and canonical solutions, by Theorem 6.2 we see that the asymmetry on the relative level y has the formula

$$\frac{|\xi^+(yA)|}{|\xi^-(yA)|} = F_y(\Theta) = F_y\left(-2\left[\frac{c^2-b}{\beta c^2-\gamma}\right]^{3/2}\frac{\lambda}{\mu}\right). \tag{6.30}$$

Therefore, the asymmetry depends on the velocity, the coefficients of linear terms b, β, γ, and on the ratio of the coefficients of nonlinear terms in micro- and macroscale $\frac{\lambda}{\mu}$. The nonlinearity coefficients μ and λ have different impact to the wave process. The nonlinearity in macroscale, i.e., μ balances the dispersion and opens the possibility for the solitary wave. But the nonlinearity in microscale, i.e., λ disturbs this balance. The ratio $\frac{\lambda}{\mu}$ affects the shape of the wave. The bigger is the quantity $-\frac{\lambda}{\mu}$, the bigger is the asymmetry. The balance between nonlinearity and dispersion collapses at the value

$$|\Theta| = 1 \quad \Leftrightarrow \quad \left|\frac{\lambda}{\mu}\right| = 2\left[\frac{\beta c^2-\gamma}{c^2-b}\right]^{3/2}.$$

Finally, we give an insight into the dependence of the asymmetry and the width of the wave on the velocity. Here we distinguish different types of dispersion of acoustic waves discussed in Chap. 4. It was shown in Sect. 4.1.1 that the cases $c_{ph} > c_g$ (normal dispersion) and $c_{ph} < c_g$ (anomalous dispersion) correspond to the relations $\frac{\gamma}{\beta} < b$ and $\frac{\gamma}{\beta} > b$, respectively. In addition, there exists a rather theoretical intermediate case $c_{ph} = c_g$ when $\frac{\gamma}{\beta} = b$ and dispersion is absent.

Let us start with the case of normal dispersion. Due to (6.10) we have

$$\frac{1}{\kappa} = (\delta\beta)^{1/2}\left[1-\frac{b-\frac{\gamma}{\beta}}{c^2-\frac{\gamma}{\beta}}\right]^{-1/2}, \qquad \Theta = -\frac{2\lambda}{\mu\beta^{3/2}}\left[1-\frac{b-\frac{\gamma}{\beta}}{c^2-\frac{\gamma}{\beta}}\right]^{3/2}. \tag{6.31}$$

Because of the inequality $\frac{\gamma}{\beta} < b$ the term $1-\frac{b-\frac{\gamma}{\beta}}{c^2-\frac{\gamma}{\beta}}$ is increasing in c^2. This implies that the width $1/\kappa$ decreases in c^2. In the case $\mu\lambda > 0$ the parameter Θ decreases in c^2. This due to (6.30) and the monotonicity of F_y (see Theorem 6.2) yields that the asymmetry is decreasing in c^2. Similarly, in the case $\mu\lambda < 0$ the asymmetry is increasing in c^2. The solvability condition $|\Theta| = 2[\frac{c^2-b}{\beta c^2-\gamma}]^{3/2}|\frac{\lambda}{\mu}| < 1$ provides the range for the velocity. The obtained range is different in the subcases $0 \le q < \frac{\gamma}{b}$, $\frac{\gamma}{b} \le q \le \beta$ and $\beta < q$, where

$$q = \left(\frac{2\lambda}{\mu}\right)^{2/3}.$$

More precisely,

$$c^2 \in \left(0, \frac{\frac{\gamma}{b}-q}{\beta-q}b\right) \cup (b,\infty) \quad \text{when } 0 \le q < \frac{\gamma}{b},$$

$$c^2 \in (b,\infty) \quad \text{when } \frac{\gamma}{b} \le q \le \beta,$$

$$c^2 \in \left(b, \frac{q-\frac{\gamma}{b}}{q-\beta}b\right) \quad \text{when } \beta < q.$$

Secondly, let us consider the case of anomalous dispersion. Then the term $1 - \frac{b-\frac{\gamma}{\beta}}{c^2-\frac{\gamma}{\beta}}$ in (6.31) is a decreasing function of c^2. This yields that the width $1/\kappa$ increases in c^2. Moreover, the asymmetry increases in c^2 when $\mu\lambda > 0$, and decreases in c^2 when $\mu\lambda < 0$. We obtain the following range for c^2:

$$c^2 \in (0,b) \cup \left(\frac{\frac{\gamma}{b}-q}{\beta-q}b, \infty\right) \quad \text{when } 0 \le q < \beta,$$

$$c^2 \in (0,b) \quad \text{when } \beta \le q \le \frac{\gamma}{b},$$

$$c^2 \in \left(\frac{q-\frac{\gamma}{b}}{q-\beta}b, b\right) \quad \text{when } \frac{\gamma}{b} < q.$$

It can be immediately seen that in all cases the size of the range depends on the ratio $\frac{|\lambda}{\mu|}$. The bigger is the ratio $\frac{|\lambda}{\mu|}$, the smaller is the range. In case $\mu > 0$ the amplitude is positive for $c^2 > b$ and negative for $c^2 < b$. Conversely, in case $\mu < 0$ the amplitude is negative for $c^2 > b$ and positive for $c^2 < b$.

Finally, we note that if the dispersion is absent (i.e. $b = \frac{\gamma}{\beta}$) then the range of c is

$$c^2 \in \mathbb{R} \setminus \{b\} \quad \text{when } q < \beta, \qquad c^2 \in \emptyset \quad \text{when } q \ge \beta.$$

6.2.4 Series Expansion of Solitary Wave

Equation (6.11) has an elementary solitary wave solution only in case $\Theta = \lambda = 0$ (formula (6.19)). If $\Theta \neq 0$ then the solution is a higher transcendental function. We are going to expand this solution into a Taylor series with respect to the parameter Θ. The truncations of this series are elementary functions that can be easily used for approximation procedures during the analysis and solution of direct and inverse problems.

We are going to construct the series for the functions $\xi^{\pm}(w)$ instead of $w(\xi)$, because this is more convenient. Due to (6.11), the derivative of $\xi(w) = \xi^{\pm}(w)$

solves the following equation with fixed w between 0 and A:

$$\xi'(w) - \frac{\Theta}{\kappa A} = \kappa^2 w^2 \left(1 - \frac{w}{A}\right)[\xi'(w)]^3. \tag{6.32}$$

Defining new variables τ and $\psi = \psi(\tau)$ by

$$\tau = \frac{\Theta}{A} w \sqrt{1 - \frac{w}{A}}, \qquad \xi'(w) = \frac{1}{\kappa w \sqrt{1 - \frac{w}{A}}} \psi\left(\frac{\Theta}{A} w \sqrt{1 - \frac{w}{A}}\right), \tag{6.33}$$

we see that (6.32) for $\xi'(w)$ is equivalent to the following cubic equation for $\psi(\tau)$:

$$\left[\psi(\tau)\right]^3 - \psi(\tau) + \tau = 0. \tag{6.34}$$

The discriminant of this equation, which equals $4 - 27\tau^2$, is different from zero for $|\tau| < \frac{2}{3\sqrt{3}}$. Therefore, using a theorem for algebraic equations with meromorphic coefficients ([58]: Chap. 6, Theorem 14.2), we conclude that (6.34) has three solutions $\psi(\tau)$. These solutions are holomorphic and differ from each other for $|\tau| < \frac{2}{3\sqrt{3}}$. Therefore, by Taylor's theorem every such solution is expandable into a series of the form

$$\psi(\tau) = \sum_{i=0}^{\infty} d_i \tau^i$$

that is uniformly convergent in every compact subset of $(-\frac{2}{3\sqrt{3}}, \frac{2}{3\sqrt{3}})$. Inserting this series into (6.34) and equating to zero the coefficients of different powers of τ we reach the following recursive formulas for d_i:

$$\begin{aligned} &d_0^3 - d_0 = 0, \qquad d_1 = \left(1 - 3d_0^2\right)^{-1}, \\ &d_i = (1 - 3d_0^2)^{-1} \sum_{\substack{0 \le i_1, i_2, i_3 < i \\ i_1 + i_2 + i_3 = i}} d_{i_1} d_{i_2} d_{i_3} \quad \text{for } i \ge 2. \end{aligned} \tag{6.35}$$

We choose the sequences starting with $d_0 = \pm 1$. (The third case $d_0 = 0$ leads to a solution that is not related to the solitary wave.) The members of the sequences of interest are

$$d_0 = \pm 1, \qquad d_1 = -\frac{1}{2}, \qquad d_2 = \mp \frac{3}{8}, \qquad d_3 = -\frac{11}{32}, \qquad \ldots.$$

Plugging the series for ψ into (6.33) we obtain

$$\xi'(w) = \frac{d_0}{\kappa}\left[w\sqrt{1 - \frac{w}{A}}\right]^{-1} + \frac{\Theta}{\kappa A} \sum_{i=0}^{\infty} d_{i+1} \Theta^i \left[\frac{w}{A}\sqrt{1 - \frac{w}{A}}\right]^i. \tag{6.36}$$

Since the convergence radius of the series $\sum_{i=0}^{\infty} d_i \tau^i$ is not less than $\frac{2}{3\sqrt{3}}$ and the relation $|\Theta| < 1$ holds for solitary waves (see Theorem 6.1), the series in the formula (6.36) is uniformly convergent for w between 0 and A. In order to obtain a formula for ξ, we must integrate (6.36) from A to w and apply the relation $\xi(A) = 0$ following from $w(0) = A$. Thus, introducing the following sequence of w (and A)—dependent functions $I_i(w) = I_i[A](w)$:

$$I_i(w) = A^{-i} \int_A^w \left[s\sqrt{1 - \frac{s}{A}} \right]^{i-1} ds$$
$$= \begin{cases} -2\ln\left[\sqrt{\frac{A}{w}}\left(1 + \sqrt{1 - \frac{w}{A}}\right)\right] & \text{if } i = 0, \\ 2\sum_{j=0}^{i-1} \binom{i-1}{j} (-1)^{j+1} \frac{(1 - \frac{w}{A})^{\frac{i+1}{2}+j}}{i+2j+1} & \text{if } i \geq 1 \end{cases} \tag{6.37}$$

we deduce the desired series expansion

$$\xi(w) = \frac{d_0}{\kappa} I_0(w) + \frac{1}{\kappa} \sum_{i=1}^{\infty} d_i \Theta^i I_i(w). \tag{6.38}$$

By means of the well-known theorem on passage to the limit under the integral it is not difficult to check that the series in (6.38) converges uniformly for any w between 0 and A.

Observing the sign of the first addend in (6.38) that dominates as $w \to 0$ we see that (6.38) with $d_0 = \pm 1$ expresses the branch $\xi(w) = \xi^{\mp}(w)$.

Moreover, the first addend

$$\xi_0(w) := \frac{d_0}{\kappa} I_0(w) = \frac{\pm I_0(w)}{\kappa}$$

corresponds to the symmetric bell-shaped solution which occurs in the case $\lambda = \Theta = 0$. On the other hand, the second addend $\frac{d_1 \Theta}{\kappa} I_1(w) = \frac{\Theta}{2\kappa}(1 - \frac{w}{A})$ is a simple linear function of Θ. The sum

$$\xi_1(w) := \frac{d_0}{\kappa} I_0(w) + \frac{\Theta}{2\kappa}\left(1 - \frac{w}{A}\right)$$

is the first order approximation of $\xi(w)$. The inverses $w_0(\xi)$ and $w_1(\xi)$ of the approximations $\xi_0(w)$ and $\xi_1(w)$, respectively, are shown in Figs. 6.9 and 6.10.

Such a series expansion is useful both in numerics and analysis. For instance, it enables the asymptotic behaviour of the solution at $w = A$ to be established. Indeed, due to the relations

$$I_0(w) \sim -\frac{2}{\sqrt{|A|}}\sqrt{|w - A|}, \qquad I_i(w) = o\big(I_{i-1}(w)\big) \quad \text{as } w \to A$$

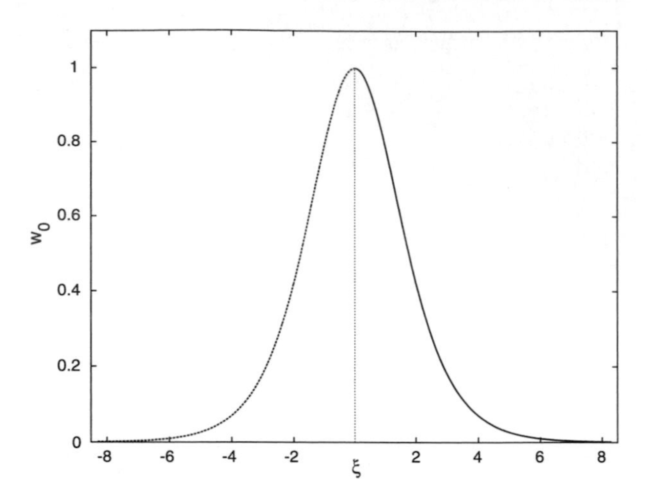

Fig. 6.9 Function $w_0(\xi)$ in case $A = \kappa = 1$

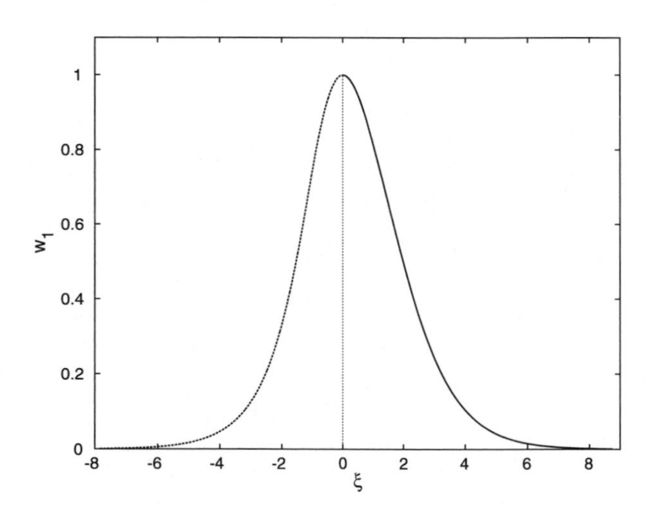

Fig. 6.10 Function $w_1(\xi)$ in case $A = \kappa = 1$, $\Theta = 0.9$

following from the definition of I_i, and the uniform convergence of the series in (6.38), we easily deduce the following asymptotic relation for the solution near the amplitude:

$$\xi(w) \sim \xi_1(w) \sim -\frac{\sqrt{|w - A|}}{|A|}\left[\frac{2d_0}{\kappa} + \frac{\Theta}{2\kappa}\sqrt{|w - A|}\right] \quad \text{as } w \to A. \tag{6.39}$$

Here $d_0 = 1$ when $\xi = \xi^-$ and $d_0 = -1$ when $\xi = \xi^+$.

6.3 Solitary Wave Solutions of Coupled System

In this section we study the system (3.40), (3.41), i.e.,

$$\left.\begin{aligned} v_{tt} &= a_0 v_{xx} + \frac{\mu}{2}(v^2)_{xx} + \vartheta_0 \varphi_{xx}, \\ \delta\varphi_{tt} &= \delta a_1 \varphi_{xx} + \delta^{3/2}\nu_1 \varphi_x \varphi_{xx} - \alpha\varphi - \vartheta_1 v. \end{aligned}\right\} \tag{6.40}$$

Travelling wave solutions of (6.40) have the form

$$v(x,t) = w(x - c_1 t), \qquad \varphi(x,t) = \chi(x - c_2 t)$$

where c_1 and c_2 are the velocities. Firstly, we note that these velocities must be equal. Indeed, from the second equation we get

$$\begin{aligned} \vartheta_1 w(x - c_1 t) = \delta \frac{\partial^2}{\partial t^2}\chi(x - c_2 t) - \delta a_1 \frac{\partial^2}{\partial x^2}\chi(x - c_2 t) \\ + \delta^{3/2}\nu_1 \frac{\partial}{\partial x}\chi(x - c_2 t)\frac{\partial^2}{\partial x^2}\chi(x - c_2 t) + \alpha\chi(x - c_2 t). \end{aligned}$$

The left and right-hand sides of this relation represent waves that propagate with the velocities c_1 and c_2, respectively, Therefore, $c_1 = c_2 = c$.

We are seeking solitary wave solutions of this system (6.40), namely the solutions of the form

$$v(x,t) = w(x - ct), \qquad \varphi(x,t) = \chi(x - ct) \tag{6.41}$$

where c is a constant velocity and the functions w, φ belong to the space $\mathcal{W}_2$ defined by

$$\mathcal{W}_2 = \left\{ w \in C^2(\mathbb{R}) : w \not\equiv 0 \text{ and } w^{(j)}(\xi) \to 0 \text{ as } |\xi| \to \infty,\ j = 0, 1 \right\}.$$

6.3.1 Separation of Unknowns. Reduction of System

Suppose that the system (6.40) has a solution of the form (6.41) where $(w, \chi) \in \mathcal{W}_2 \times \mathcal{W}_2$. Simple substitution brings this system to the form

$$\left.\begin{aligned} c^2 w''(\xi) &= a_0 w''(\xi) + \frac{\mu}{2}[w^2(\xi)]'' + \vartheta_0 \chi''(\xi), \\ \delta c^2 \chi''(\xi) &= \delta a_1 \chi''(\xi) + \delta^{3/2}\nu_1 \chi'(\xi)\chi''(\xi) - \alpha\chi(\xi) - \vartheta_1 w(\xi). \end{aligned}\right\} \tag{6.42}$$

We are going to separate the quantities w and χ. To this end, from the first equation we express χ'':

$$\chi''(\xi) = \frac{1}{\vartheta_0}\left\{ \left(c^2 - a_0\right) w''(\xi) - \frac{\mu}{2}\left[w^2(\xi)\right]'' \right\}$$

and integrate twice. In view of the vanishing conditions for χ, χ', w and w' at infinity we obtain the formula for χ in terms of w:

$$\chi(\xi) = \frac{1}{\vartheta_0}\left\{(c^2 - a_0)w(\xi) - \frac{\mu}{2}w^2(\xi)\right\}. \tag{6.43}$$

Further, let us substitute χ by (6.43) in the second equation of (6.42). Multiplicating the resulting expression by ϑ_0 and observing the relations

$$\vartheta = \vartheta_0\vartheta_1, \qquad \nu = \frac{\nu_1}{\vartheta_0}$$

we arrive at the following single equation for w:

$$\begin{aligned}&\left[\delta(c^2 - a_1) - \delta^{3/2}\nu\left\{(c^2 - a_0)w - \frac{\mu}{2}w^2\right\}'\right]\left\{(c^2 - a_0)w - \frac{\mu}{2}w^2\right\}'' \\ &\quad + \alpha\left\{(c^2 - a_0)w - \frac{\mu}{2}w^2\right\} + \vartheta w = 0.\end{aligned} \tag{6.44}$$

We continue studying solitary wave solutions of this equation.

Lemma 6.3 *Let* (6.44) *have a solution in the space* $\mathcal{W}_2$. *Then* $c^2 - a_1 \neq 0$, $c^2 - a_0 \neq 0$ *and* $\mu \neq 0$. *Moreover,* $(c^2 - a_0)w'(\xi) - \frac{\mu}{2}[w^2(\xi)]' \neq 0$ *a.e.*, $c^2 - a_0 - \mu w(\xi) \neq 0$ *a.e. and*

$$\delta(c^2 - a_1) - \delta^{3/2}\nu\left\{(c^2 - a_0)w'(\xi) - \frac{\mu}{2}\left[w^2(\xi)\right]'\right\} \neq 0 \quad a.e.$$

The proof is included in Sect. 6.4.

Let us continue the reduction of (6.44). We multiply it by $(c^2 - a_0)w' - \frac{\mu}{2}[w^2]'$ to get

$$\begin{aligned}&\left[\delta(c^2 - a_1)\left\{(c^2 - a_0)w - \frac{\mu}{2}w^2\right\}' - \delta^{3/2}\nu\left[\left\{(c^2 - a_0)w - \frac{\mu}{2}w^2\right\}'\right]^2\right] \\ &\quad \times \left\{(c^2 - a_0)w - \frac{\mu}{2}w^2\right\}'' \\ &\quad + \alpha\left\{(c^2 - a_0)w - \frac{\mu}{2}w^2\right\}\left\{(c^2 - a_0)w - \frac{\mu}{2}w^2\right\}' \\ &\quad + \vartheta w\{(c^2 - a_0)w' - \mu w w'\} = 0.\end{aligned} \tag{6.45}$$

Due to the non-vanishing assertion about the function $(c^2 - a_0)w' - \frac{\mu}{2}[w^2]'$ in Lemma 6.3, (6.44) and (6.45) are equivalent in the space $\mathcal{W}_2$. Integrating once again, (6.45) is reduced to the following equivalent equation of the first order:

$$\frac{\delta(c^2-a_1)}{2}\{(c^2-a_0)w'-\mu ww'\}^2-\frac{\delta^{3/2}\nu}{3}\{(c^2-a_0)w'-\mu ww'\}^3$$
$$=-\frac{\alpha}{2}\left\{(c^2-a_0)w-\frac{\mu}{2}w^2\right\}^2-\frac{\vartheta(c^2-a_0)}{2}w^2+\frac{\vartheta\mu}{3}w^3. \tag{6.46}$$

Before proving the existence theorem for (6.46), let us transform it to a canonical form. Firstly, we divide by $\delta(c^2-a_1)(c^2-a_0)^2/2\neq 0$ to get

$$\left\{\left(1-\frac{\mu w}{c^2-a_0}\right)w'\right\}^2-\frac{2\delta^{1/2}\nu(c^2-a_0)}{3(c^2-a_1)}\left\{\left(1-\frac{\mu w}{c^2-a_0}\right)w'\right\}^3=P(w) \tag{6.47}$$

where

$$P(w)=w^2\left[\frac{a_0\alpha-c^2\alpha-\vartheta}{\delta(c^2-a_1)(c^2-a_0)}+\frac{\mu(2\vartheta+3c^2\alpha-3a_0\alpha)}{3\delta(c^2-a_1)(c^2-a_0)^2}w\right.$$
$$\left.-\frac{\mu^2\alpha}{4\delta(c^2-a_1)(c^2-a_0)^2}w^2\right]. \tag{6.48}$$

Lemma 6.4 *Let* (6.44) *have a solution in the space* $\mathcal{W}_2$. *Then*

$$\kappa^2:=\frac{a_0\alpha-c^2\alpha-\vartheta}{\delta(c^2-a_1)(c^2-a_0)}>0 \tag{6.49}$$

and

$$p\in(-\infty,-1)\cup\left(\frac{1}{3},\infty\right)\quad \textit{where } p=\frac{\vartheta}{3(c^2\alpha-a_0\alpha+\vartheta)}. \tag{6.50}$$

The proof is shifted to Sect. 6.4, again.

Since $\kappa\neq 0$, we can bring κ^2 out as a factor of $P(w)$. According to (6.50), $P(w)$ has two single real roots that are different from 0. More precisely, we can factorise $P(w)$ as follows.

$$P(w)=\kappa^2w^2\left[1-\frac{\mu}{c^2-a_0}(1-p)w+\frac{\mu^2}{4(c^2-a_0)^2}(1-3p)w^2\right]$$
$$=\kappa^2w^2\left[1-\frac{\mu}{c^2-a_0}\frac{1-p-\sqrt{p+p^2}}{2}w\right]$$
$$\times\left[1-\frac{\mu}{c^2-a_0}\frac{1-p+\sqrt{p+p^2}}{2}w\right]. \tag{6.51}$$

Further, let us introduce the following parameters:

$$\Theta_1=\begin{cases}\frac{2}{1-p+\sqrt{p+p^2}} & \text{in case } p\in(-\infty,-1),\\ \frac{2}{1-p-\sqrt{p+p^2}} & \text{in case } p\in(\frac{1}{3},\infty),\end{cases} \tag{6.52}$$

$$\Theta_2 = \begin{cases} \frac{2}{1-p-\sqrt{p+p^2}} & \text{in case } p \in (-\infty, -1), \\ \frac{2}{1-p+\sqrt{p+p^2}} & \text{in case } p \in (\frac{1}{3}, \infty). \end{cases} \tag{6.53}$$

They are put together from the inverses of the p-dependent coefficients $\frac{1-p\pm\sqrt{p+p^2}}{2}$ of (6.51) in such a way that the relation

$$\frac{\Theta_2}{\Theta_1} \notin [0, 1] \tag{6.54}$$

is achieved. The meaning of this relation will become clear below, in the existence proofs. By means of straightforward computations it is possible to show that the formula

$$\Theta_2 = \frac{3\Theta_1 - 4}{2\Theta_1 - 3} \tag{6.55}$$

is valid.

Since $p \in (-\infty, -1) \cup (\frac{1}{3}, \infty)$, the ranges of Θ_1 and Θ_2 are

$$\Theta_1 \in (-\infty, 0) \cup (0, 1) \quad \text{and} \quad \Theta_2 \in \left(1, \frac{4}{3}\right) \cup \left(\frac{4}{3}, \frac{3}{2}\right), \tag{6.56}$$

respectively.

Moreover, let us define the additional parameters

$$\begin{aligned} A_0 &= \frac{c^2 - a_0}{\mu}, \\ \Theta &= \kappa A_0 \frac{2\delta^{1/2}\nu(c^2 - a_0)}{3(c^2 - a_1)} = \frac{2}{3}\frac{\nu}{\mu}\frac{(c^2 - a_0)^2}{c^2 - a_1}\sqrt{\frac{a_0\alpha - c^2\alpha - \vartheta}{(c^2 - a_0)(c^2 - a_1)}}. \end{aligned} \tag{6.57}$$

Summing up, by virtue of the relations (6.51), (6.52), (6.55) and (6.57), the equation (6.47) can be rewritten in the form

$$\begin{aligned} &\left\{\left(1 - \frac{w}{A_0}\right)w'\right\}^2 - \frac{\Theta}{\kappa A_0}\left\{\left(1 - \frac{w}{A_0}\right)w'\right\}^3 \\ &\quad = \kappa^2 w^2\left(1 - \frac{w}{A_0\Theta_1}\right)\left(1 - \frac{w}{A_0\Theta_2}\right). \end{aligned} \tag{6.58}$$

Due to the non-vanishing assertions of Lemma 6.3, the equation (6.58) possesses the equivalent form of the autonomous system, too:

$$\begin{aligned} w' &= W, \\ W' &= \frac{\mu W^2\{\delta(c^2 - a_1) - \delta^{3/2}\nu(c^2 - a_0 - \mu w)W\} - \alpha\{(c^2 - a_0)w - \frac{\mu}{2}w^2\} - \vartheta w}{\{\delta(c^2 - a_1) - \delta^{3/2}\nu(c^2 - a_0 - \mu w)W\}(c^2 - a_0 - \mu w)}. \end{aligned} \tag{6.59}$$

The introduced parameters have definite geometrical meanings. Due to the conditions $w, w' \to 0$ as $|\xi| \to \infty$ in $\mathcal{W}_2$, from (6.47), (6.48) it follows that κ is the exponential decay rate, i.e., the asymptotic formulas (6.13) and (6.14) are valid. The parameters Θ_1 and A_0 affect the shape and size of the wave: nonlinear and linear (proportional) magnification, respectively. The product $A = A_0\Theta_1$ is the amplitude of the wave. As in the case of the hierarchical equation, Θ is related to the asymmetry. The parameter Θ_2 is not free, it depends on Θ_1 (formula (6.55)).

By linear scaling we can reduce (6.58) to a canonical form that contains only two parameters. The most natural scaling seems to be

$$y = \frac{1}{A_0} w, \qquad \zeta = \kappa\xi. \tag{6.60}$$

Then the equation for the new unknown y can be written

$$\{(1-y)y'\}^2 - \Theta\{(1-y)y'\}^3 = y^2\left(1 - \frac{y}{\Theta_1}\right)\left(1 - \frac{y}{\Theta_2}\right). \tag{6.61}$$

This is an equation for a wave with the amplitude Θ_1 and unit exponential decay rate.

6.3.2 *Existence and Basic Properties of Canonical Waves*

In order to simplify the study of the existence, we have to perform additional nonlinear "scaling" of (6.61). The reason is that the critical lines of the autonomous system corresponding to (6.61) are 3rd order curves. Such a circumstance makes the treatment of this equation cumbersome. Our idea is to straighten the critical lines. To this end, let us define the new variables

$$\eta(\zeta) = \int \frac{d\zeta}{1 - y(\zeta)} \quad \text{with } \eta(0) = 0 \quad \text{and} \quad \hat{y}(\eta) = \frac{1}{\Theta_1} y(\zeta). \tag{6.62}$$

Note that the inverse formulas that transform η to ζ and $\hat{y}$ to y are

$$\zeta(\eta) = \int \left(1 - \Theta_1\hat{y}(\eta)\right) d\eta \quad \text{with } \zeta(0) = 0 \quad \text{and} \quad y(\zeta) = \Theta_1\hat{y}(\eta). \tag{6.63}$$

The following lemma whose proof is shifted to Sect. 6.4 transforms the equation for y to a constrained equation for $\hat{y}$.

Lemma 6.5 *If $y \in \mathcal{W}_2$ solves* (6.61) *then there exist $M_2 > M_1 > 0$ such that η and $\hat{y}$ defined by* (6.62) *satisfy $\eta'(\zeta) \in [M_1, M_2]$, $\hat{y} \in \mathcal{W}_2$ and $\hat{y}$ solves the following problem*:

$$(\hat{y}')^2 - \Theta\Theta_1(\hat{y}')^3 = \hat{y}^2(1-\hat{y})\left(1 - \frac{\Theta_1}{\Theta_2}\hat{y}\right), \qquad 1 - \Theta_1\hat{y} > 0. \tag{6.64}$$

Conversely, if $\hat{y} \in \mathcal{W}_2$ *solves* (6.64) *then there exist* $\hat{M}_2 > \hat{M}_1 > 0$ *such that* ζ *and* y *defined by* (6.63) *satisfy* $\zeta'(\eta) \in [\hat{M}_1, \hat{M}_2]$, $y \in \mathcal{W}_2$ *and* y *solves* (6.61).

The equivalent differentiated form of (6.64) is

$$(2 - 3\Theta\Theta_1\hat{y}')\hat{y}'\hat{y}'' = 2\hat{y}(1 - \Theta_1\hat{y})\left(1 - \frac{2}{\Theta_2}\hat{y}\right)\hat{y}', \qquad 1 - \Theta_1\hat{y} > 0. \tag{6.65}$$

In the factorisation of the right-hand side of (6.65) we used the relation

$$3 + \frac{3\Theta_1}{\Theta_2} = 2\Theta_1 + \frac{4}{\Theta_2}$$

following from (6.55). The quantities $\hat{y}'$ and $2 - 3\Theta\Theta_1\hat{y}'$ in (6.65) can be represented in terms of w:

$$\hat{y}' = \frac{1}{\Theta_1}(1 - y)y' = \frac{1}{\kappa A}\left[(c^2 - a_0)w' - \frac{\mu}{2}\left[w'\right]^2\right],$$

$$2 - 3\Theta\Theta_1\hat{y}' = \frac{2}{\delta(c^2 - a_1)}\left[\delta(c^2 - a_1) - \delta^{3/2}\nu\left\{(c^2 - a_0)w'(\xi) - \frac{\mu}{2}\left[w^2(\xi)\right]'\right\}\right].$$

Applying Lemma 6.3 we obtain $\hat{y}(\eta) \neq 0$ a.e. and $2 - 3\Theta\Theta_1\hat{y}'(\eta) \neq 0$ a.e. Therefore, we can divide (6.65) by $\hat{y}'(2 - 3\Theta\Theta_1\hat{y}')$ and rewrite it in the form of the constrained first order system:

$$\hat{y}' = z, \qquad z' = \frac{2\hat{y}(1 - \Theta_1\hat{y})(1 - \frac{2}{\Theta_2}\hat{y})}{2 - 3\Theta\Theta_1 z}, \qquad 1 - \Theta_1\hat{y} > 0. \tag{6.66}$$

The critical lines of the system (6.66) in the phase plane are $z = \frac{2}{3\Theta\Theta_1}$ (the singularity line) and $\hat{y} = 0$, $\hat{y} = \frac{1}{\Theta_1}$, $\hat{y} = \frac{\Theta_2}{2}$ (the zero lines).

Firstly, let us study the case $\Theta\Theta_1 > 0$. Depending on the sign of Θ_1, the phase portrait (orientation of the vector of the right-hand side of the system (6.66)) can be of two kinds (Figs. 6.11 and 6.12). From these figures we see that the trajectory T satisfying (6.21) can only be located in

$$\begin{aligned} &\left\{(\hat{y}, z) : 0 \leq \hat{y} < \frac{1}{\Theta_1}, z < \frac{2}{3\Theta\Theta_1}\right\} \quad \text{if } 0 < \Theta_1 < 1, \\ &\left\{(\hat{y}, z) : \hat{y} \geq 0, z < \frac{2}{3\Theta\Theta_1}\right\} \qquad\quad \text{if } \Theta_1 < 0. \end{aligned} \tag{6.67}$$

Note that the line $\hat{y} = \frac{\Theta_2}{2}$ is always located on the right half-plane $\hat{y} > 0$ and in case $0 < \Theta_1 < 1$ to the left of the line $\hat{y} = \frac{1}{\Theta_1}$ (cf. (6.56)).

The existence proof for (6.64) is similar to the existence proof for (6.16) in the previous section. We start by introducing the equation of the trajectory T which is

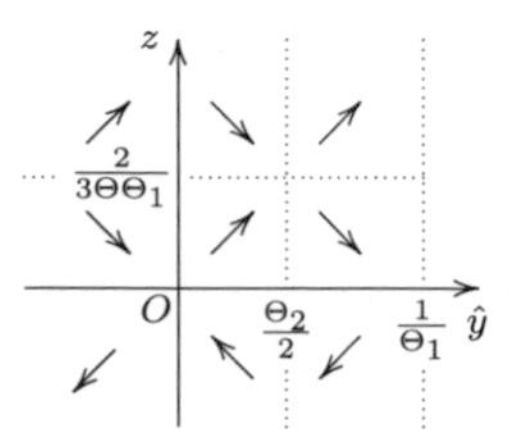

Fig. 6.11 Phase portrait if $0 < \Theta_1 < 1$

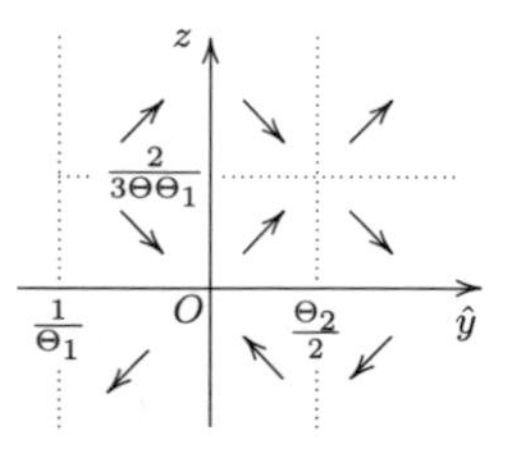

Fig. 6.12 Phase portrait if $\Theta_1 < 0$

$z^2 - \Theta\Theta_1 z^3 = \hat{y}^2(1-\hat{y})(1-\frac{\Theta_1}{\Theta_2}\hat{y})$ and continue to study this equation in the subdomains (6.67). To this end we express it as $z = f^{-1}[\hat{y}^2(1-\hat{y})(1-\frac{\Theta_1}{\Theta_2}\hat{y})]$ where f^{-1} is the inverse of $f(z) = z^2 - \Theta\Theta_1 z^3$ and split it up:

$$z = f^{-1}(g), \qquad g = \hat{y}^2(1-\hat{y})\left(1 - \frac{\Theta_1}{\Theta_2}\hat{y}\right).$$

Observing the relations (6.54) and (6.56) it is easy to see that the function $g(\hat{y})$ has the shape graphed in Figs. 6.14 and 6.15. On the other hand, the function f^{-1} depicted in Fig. 6.13 has three branches: f_1^{-1}, f_2^{-1} and f_3^{-1}. The latter is excluded because it falls beyond the singularity line $z = \frac{2}{3\Theta\Theta_1}$. The remaining branches f_1^{-1} and f_2^{-1} are defined for nonnegative values of g. This together with the inequalities for $\hat{y}$ in (6.67) restricts the domain of g to [0, 1]. (Here we also take the inequality

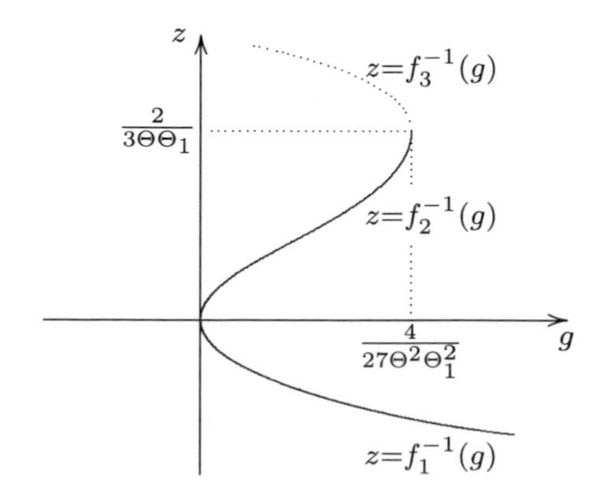

Fig. 6.13 Function $f^{-1}(g)$

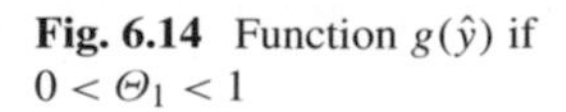

Fig. 6.14 Function $g(\hat{y})$ if $0 < \Theta_1 < 1$

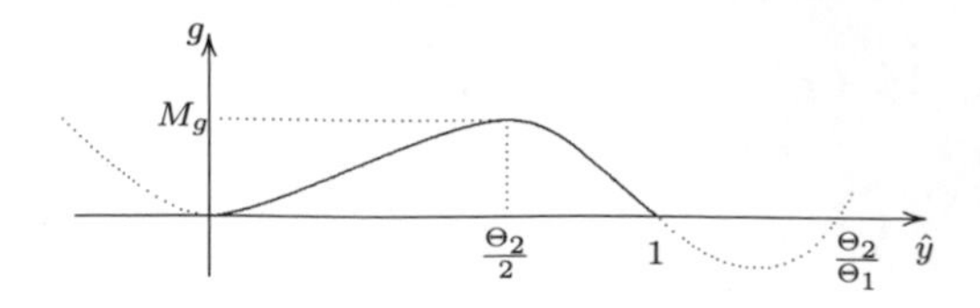

Fig. 6.15 Function $g(\hat{y})$ if $\Theta_1 < 0$

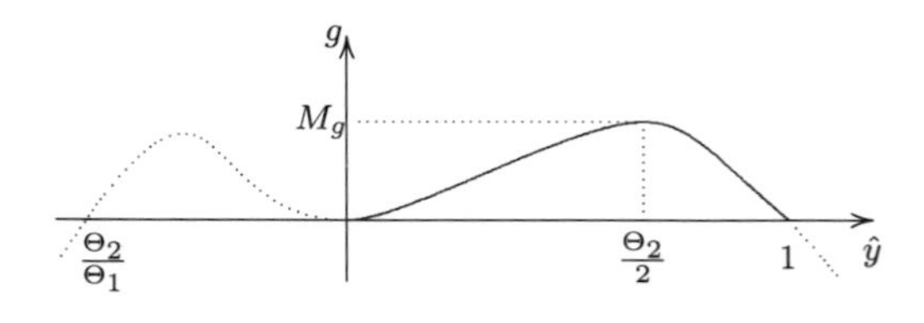

$\frac{1}{\Theta_1} < \frac{\Theta_2}{\Theta_1}$ into account if $0 < \Theta_1 < 1$.) The maximum of g in the interval $[0, 1]$ is

$$M_g = g\left(\frac{\Theta_2}{2}\right) = \frac{\Theta_2^2}{4}\left(1 - \frac{\Theta_1}{2}\right)\left(1 - \frac{\Theta_2}{2}\right).$$

The branch f_1^{-1} forms the curve $z = f_1^{-1}[g(\hat{y})]$, which connects the points $(0, 0)$ and $(1, 0)$ and passes through the lower half-plane $z < 0$. To study the branch f_2^{-1}, let us introduce the quantity

$$\mathcal{D} = \Theta^2\Theta_1^2\Theta_2^2\left(1 - \frac{\Theta_1}{2}\right)\left(1 - \frac{\Theta_2}{2}\right).$$

By (6.56), $\mathcal{D} \geq 0$. We are going to treat separately the cases $\mathcal{D} > \frac{16}{27}$, $\mathcal{D} = \frac{16}{27}$ and $0 \leq \mathcal{D} < \frac{16}{27}$.

(1) $\mathcal{D} > \frac{16}{27}$. Comparing the range $[0, M_g]$ of g with the domain $[0, \frac{4}{27\Theta^2\Theta_1^2}]$ of f_2^{-1}, we see that the whole range of g extends beyond the domain of f_2^{-1}. Restricting the range of g to the interval $[0, \frac{4}{27\Theta^2\Theta_1^2}]$ restricts the domain of g to a union $[0, y_1] \cup [y_2, 1]$ with some $y_1 < y_2$. Thus, the composition $z = f_2^{-1}[g(y)]$ is not defined for $y \in (y_1, y_2)$ (see Fig. 6.16). The system (6.66) has no trajectory with property (6.21). The solitary wave solution does not exist in $\mathcal{W}_2$.

(2) $\mathcal{D} = \frac{16}{27}$. Then the curve $z = f_2^{-1}[g(\hat{y})]$ connects the points $(0, 0)$ and $(1, 0)$ and passes through the upper half-plane $z > 0$. The maximum of $z = f_2^{-1}[g(\hat{y})]$ is achieved at the point $(\hat{y}_*, z_*) = (\frac{\Theta_2}{2}, \frac{2}{3\Theta\Theta_1})$ on the singularity line. Since we are seeking the solutions $\hat{y} \in \mathcal{W}_2$, the trajectory T must be a smooth curve. Therefore, the necessary extremum condition $\frac{dz}{d\hat{y}} = 0$ must be valid at $(\hat{y}_*, z_*)$. Let us plug $z = z_* + \Delta z$ and $\hat{y} = \hat{y}_* + \Delta\hat{y}$ into the equation $z^2 - \Theta\Theta_1 z^3 = \hat{y}^2(1 - \hat{y})(1 - \frac{\Theta_1}{\Theta_2}\hat{y})$ and expand with respect to Δz and $\Delta\hat{y}$. Simplifying the

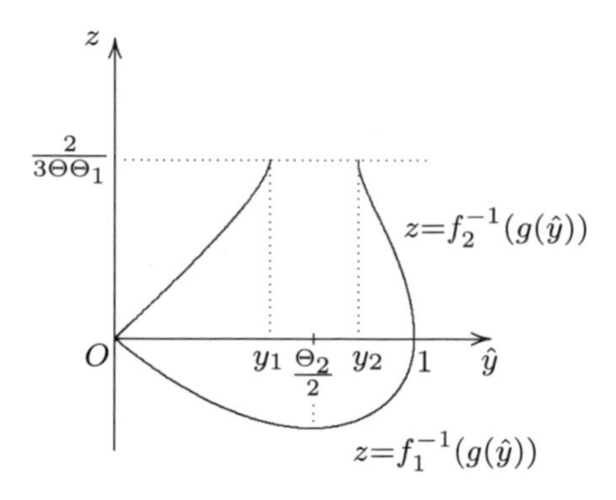

Fig. 6.16 T in case $\mathcal{D} > \frac{16}{27}$

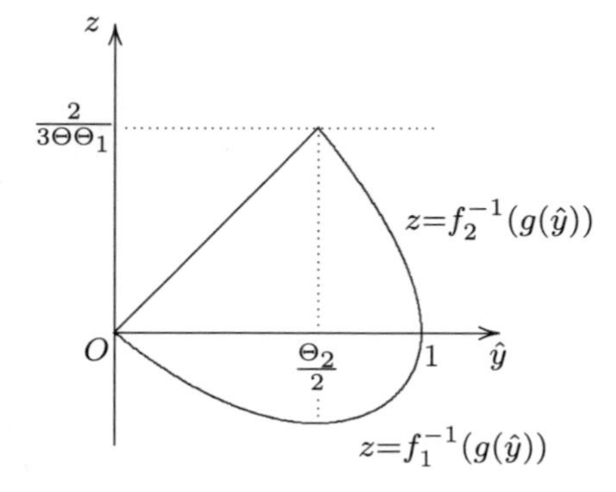

Fig. 6.17 T in case $\mathcal{D} = \frac{16}{27}$

resulting expression by means of the relations (6.55) and $\mathcal{D} = \frac{16}{27}$ we obtain

$$\begin{aligned}&\frac{1}{3}\Delta z^2 - \Theta\Theta_1\Delta z^3\\&\quad= \frac{1}{2}(\Theta_1\Theta_2 - 2)\Delta\hat{y}^2 + \frac{4}{3\Theta_2}(\Theta_1\Theta_2 - 1)\Delta\hat{y}^3 + \frac{\Theta_1}{\Theta_2}\Delta\hat{y}^4.\end{aligned}$$

Thus, $|\frac{dz}{d\hat{y}}| = \sqrt{\frac{3}{2}(\Theta_1\Theta_2 - 2)} \neq 0$ at $(\hat{y}_*, z_*)$. We have reached a contradiction. This means that T is not smooth (see Fig. 6.17). Equation (6.64) does not have a solution in $\mathcal{W}_2$.

(3) $0 \leq \mathcal{D} < \frac{16}{27}$. Then $M_g < \frac{4}{27\Theta^2\Theta_1^2}$. The curve $z = f_2^{-1}[g(\hat{y})]$ connects the points $(0,0)$ and $(1,0)$ and passes through the band $0 < z < \frac{2}{3\Theta\Theta_1}$. This means that the trajectory T, i.e., the union of the curves $z = f_1^{-1}[g(\hat{y})]$ and $z = f_2^{-1}[g(\hat{y})]$, possesses the property (6.21) (see Fig. 6.18). The Cauchy problem for (6.66) with the conditions $\hat{y}(0) = 1$, $z(0) = 0$ has a unique solution, which by the relation $z^2 \sim \hat{y}^2$ as $\hat{y} \to 0$ (cf. (6.64)) satisfies the conditions $\hat{y}(\eta), z(\eta) = \hat{y}'(\eta) \to 0$ as $|\eta| \to \infty$. Consequently, the function $\hat{y}(\eta)$ that is defined for $\eta \in \mathbb{R}$ solves (6.64). Furthermore, since the right-hand side of (6.66) is infinitely differentiable for $\hat{y} \geq 0, z < \frac{2}{3\Theta\Theta_1}$, the solution $\hat{y}(\eta)$ is also infinitely differentiable.

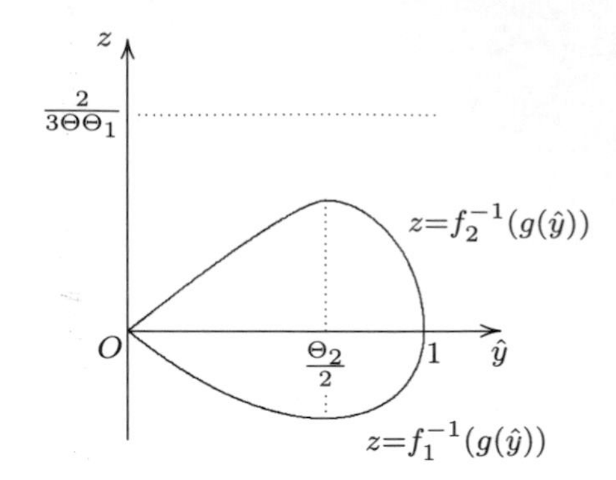

Fig. 6.18 T in case $0 \le \mathcal{D} < \frac{16}{27}$

Therefore, $\hat{y}(\eta)$ belongs to $\mathcal{W}_2$. All other solutions of (6.64) in $\mathcal{W}_2$ can be derived from $\hat{y}(\eta)$ by the argument shift $\eta \mapsto \eta + C$, where C is a constant.

Evidently, $\hat{y}(\eta)$ solves (6.64) if and only if $\hat{y}(-\eta)$ solves (6.64) with Θ replaced by $-\Theta$. Therefore, any solution corresponding to $\Theta\Theta_1 < 0$ is the reflection over the line $\eta = 0$ of the solution corresponding to $\Theta\Theta_1 > 0$.

Let's sum up these arguments. We have proved that the constrained equation (6.64) has a solution in the space $\mathcal{W}_2$ if and only if

$$\Theta^2\Theta_1^2\Theta_2^2\left(1-\frac{\Theta_1}{2}\right)\left(1-\frac{\Theta_2}{2}\right) < \frac{16}{27}. \tag{6.68}$$

Moreover, the set of all solutions of (6.64) has the form $\{\hat{y}_C(\eta) = \hat{y}_0(\eta+C) : C \in \mathbb{R}\}$ where $\hat{y} = \hat{y}_0 \in \mathcal{W}_2$ is the unique solution that satisfies the conditions $\hat{y}_0(0) = 1$, $\hat{y}'(0) = 0$. From the qualitative behaviour of T we immediately deduce the following properties for $\hat{y} = \hat{y}_0$:

$$\left.\begin{array}{l}
\hat{y} \text{ is positive and has the maximal value 1 at } \eta = 0;\\
\hat{y} \text{ increases for } \eta < 0 \text{ and decreases for } \eta > 0;\\
2 - 3\Theta\Theta_1\hat{y}' > 0;\\
\text{there exist } \eta_1 < 0 \text{ and } \eta_2 > 0 \text{ such that } \hat{y} \text{ is}\\
\quad \text{concave for } \eta < \eta_1, \eta > \eta_2 \text{ and convex for } \eta_1 < \eta < \eta_2;\\
\hat{y}(\eta_1) = \hat{y}(\eta_2) = \dfrac{\Theta_2}{2}.
\end{array}\right\} \tag{6.69}$$

Next we consider the canonical wave equation (6.61). Due to Lemma 6.5 that states the equivalence of (6.64) and (6.61), the proved existence assertion automatically holds for (6.61), too. The qualitative behaviour of y and its derivatives can easily be deduced from (6.69), the expressions

$$\Theta^{-1}y = \hat{y}, \qquad \Theta_1^{-1}y' = \frac{\hat{y}'}{1-\Theta_1\hat{y}},$$

$$\Theta_1^{-1} y^{**} = \frac{\hat{y}''}{(1-\Theta_1\hat{y})^2} \quad \text{for } y^{**} = y'' - \frac{(y')^2}{1-y}$$

following from (6.62) and (6.63) and the relation $1-\Theta_1\hat{y} > 0$. The results for (6.61) are summarised in the following theorem.

Theorem 6.4 *The canonical equation* (6.61) *has solutions in* $\mathcal{W}_2$ *if and only if* (6.68) *is satisfied. The set of all solutions of this problem in* $\mathcal{W}_2$ *has the form* $\{y_C(\zeta) = y_0(\zeta + C) : C \in \mathbb{R}\}$, *where* $y = y_0 \in \mathcal{W}_2$ *is the unique solution satisfying the following properties*:

$$\left.\begin{array}{l}
\Theta_1^{-1} y \text{ is positive and has the maximal value } 1 \text{ at } \zeta = 0;\\
\Theta_1^{-1} y \text{ increases for } \zeta < 0 \text{ and decreases for } \zeta > 0;\\
2 - 3\Theta(1-y)y' > 0;\\
\text{there exist } \zeta_1 < 0 \text{ and } \zeta_2 > 0 \text{ such that}\\
\Theta_1^{-1} y^{**} > 0 \text{ for } \zeta < \zeta_1, \zeta > \zeta_2,\ \Theta_1^{-1} y^{**} < 0 \text{ for } \zeta_1 < \zeta < \zeta_2;\\
y(\zeta_1) = y(\zeta_2) = \dfrac{\Theta_1\Theta_2}{2}.
\end{array}\right\} \tag{6.70}$$

Clearly, Θ_1 is the amplitude of the wave $y(\zeta)$. But Θ_1 has another meaning, too. It is related to the steepness of the wave. Another free parameter Θ is related to the asymmetry of the wave. To prove these statements, we first define

$$\zeta = \zeta^-(y) < 0 \quad \text{and} \quad \zeta = \zeta^+(y) > 0$$

—the left and right inverses of the function $y = y(\zeta)$, respectively.

The quantities $\zeta^{\pm}(y)$ depend on Θ_1 and Θ.

Theorem 6.5

(i) *The quantities* $|\zeta^{\pm}|'(y)$ *are increasing with respect to* Θ_1 *for any fixed* Θ *and* y.

(ii) *The equality*

$$\frac{|\zeta^+(y)|}{|\zeta^-(y)|} = \bar{F}_{\Theta_1,y}(\Theta) \tag{6.71}$$

is valid for any fixed Θ_1 *and* y *where* $\bar{F}_{\Theta_1,y}$ *is a function of* Θ *in the interval* $(-b_{\Theta_1}, b_{\Theta_1})$ *with*

$$b_{\Theta_1} = 4\left[27\Theta_1^2\Theta_2^2\left(1-\frac{\Theta_1}{2}\right)\left(1-\frac{\Theta_2}{2}\right)\right]^{-1/2}$$

possessing the following properties: $\bar{F}_{\Theta_1,y}(\Theta)$ *increases in case* $0 < \Theta_1 < 1$ *and decreases in case* $\Theta_1 < 0$ *and* $\bar{F}_{\Theta_1,y}(0) = 1$.

The proof is included in Sect. 6.4.

6.3.3 Properties of General Solitary Waves

Now let us consider (6.58) depending on the parameters κ, A_0, Θ_1 and Θ. Due to Theorem 6.4, the equation possesses a solitary wave solution in $\mathcal{W}_2$ if and only if the inequality (6.68) is satisfied. The solution has the form $w(\xi) = A_0 y_{\Theta,\Theta_1}(\kappa\xi)$, where y_{Θ,Θ_1} is the solution of (6.61) corresponding to the parameters Θ and Θ_1 (recall that Θ_2 depends on Θ_1).

The existence conditions are in the case of the coupled system more restrictive than in the case of the hierarchical equation. Indeed, in addition to the positivity condition for κ^2 (formula (6.49)) and the nonlinearity restriction (6.68), the discriminant condition (6.50) must be satisfied. Note that in the case of the hierarchical equation such a condition didn't occur. This phenomenon can be explained as follows. For the existence, the polynomial of w on right-hand side of the solitary wave equation (i.e. (6.9) and (6.47)) must have at least one non-zero real root. This implies discriminant restrictions for polynomials of even order as in (6.47), but not for polynomials of odd order as in (6.9).

Observing the positivity of α and ϑ, the discriminant condition (6.50) can be rewritten in terms of $\frac{\alpha}{\vartheta}(a_0 - c^2)$ and c^2 as follows:

$$\begin{aligned} &\text{either} \quad \frac{\alpha}{\vartheta}\left(a_0 - c^2\right) \in \left(1, \frac{4}{3}\right) \quad \Leftrightarrow \quad c^2 \in \left(a_0 - \frac{4\vartheta}{3\alpha}, a_0 - \frac{\vartheta}{\alpha}\right) \quad \text{(case I)} \\ &\text{or} \quad \frac{\alpha}{\vartheta}\left(a_0 - c^2\right) \in (0, 1) \quad \Leftrightarrow \quad c^2 \in \left(a_0 - \frac{\vartheta}{\alpha}, a_0\right) \quad \text{(case II)}. \end{aligned} \tag{6.72}$$

In the cases I and II we have $p \in (-\infty, -1)$, $\Theta_1 \in (0, 1)$ and $p \in (\frac{1}{3}, \infty)$, $\Theta_1 \in (-\infty, 0)$, respectively. The product

$$A = A_0\Theta_1 = \frac{c^2 - a_0}{\mu}\Theta_1, \tag{6.73}$$

which is the amplitude of the wave, becomes positive in case $\frac{\Theta_1}{\mu} < 0$ and negative in case $\frac{\Theta_1}{\mu} > 0$. (Always $c^2 - a_0 < 0$ due to (6.72)!) The positivity condition (6.49) has the following equivalent form:

$$\frac{c^2\alpha - a_0\alpha + \vartheta}{c^2 - a_1} > 0. \tag{6.74}$$

This, in view of (6.72) can be transformed to

$$c^2 < a_1 \quad \text{in case I,} \qquad c^2 > a_1 \quad \text{in case II.} \tag{6.75}$$

The nonlinearity restriction (6.68) in terms of original coefficients can be rewritten as

$$\left(\frac{c^2 - a_1}{c^2 - a_0 + \frac{\vartheta}{\alpha}}\right)^3 > 4\vartheta \frac{\nu^2}{\mu^2}. \tag{6.76}$$

This is a stronger condition than (6.74).

Observing the definitions of A and $\Theta, \Theta_1, \Theta_2$ and taking the recent discussion into account, the results concerning the canonical equation (6.61) are easily reformulated for the general equation (6.58).

Theorem 6.6 *Let* (6.72) *and* (6.75) *be valid. Equation* (6.58) (*or, equivalently,* (6.44)) *has solutions in* $\mathcal{W}_2$ *if and only if the inequality* (6.76) *is satisfied. The set of all solutions in* $\mathcal{W}_2$ *has the form* $\{w_C(\xi) = w_0(\xi + C) : C \in \mathbb{R}\}$ *where* $w = w_0 \in \mathcal{W}_2$ *is the unique solution satisfying the following properties*:

$$\left.\begin{array}{l}
A^{-1}w \text{ is positive and has the maximal value } 1 \text{ at } \xi = 0;\\[1ex]
A^{-1}w \text{ increases for } \xi < 0 \text{ and decreases for } \xi > 0;\\[1ex]
1 - \dfrac{\delta^{1/2}\nu}{c^2 - a_1}(c^2 - a_0 - \mu w)w' > 0;\\[1ex]
\text{there exist } \xi_1 < 0 \text{ and } \xi_2 > 0 \text{ such that the function}\\[1ex]
w^{**} = w'' - \dfrac{\mu(w')^2}{c^2 - a_0 - \mu w} \text{ satisfies}\\[1ex]
A^{-1}w^{**} > 0 \text{ for } \xi < \xi_1, \xi > \xi_2,\ A^{-1}w^{**} < 0 \text{ for } \xi_1 < \xi < \xi_2;\\[1ex]
w(\xi_1) = w(\xi_2) = \dfrac{2(c^2 - a_0 + \frac{\vartheta}{\alpha})}{\mu}.
\end{array}\right\} \tag{6.77}$$

Next let us perform some shape and asymmetry analysis on the basis of Theorem 6.5. Since the function $w(\xi)$ is strictly monotone to the left and right of the extremum point $\xi = 0$, it has two inverses. Let us define

$$\xi = \xi^-(w) < 0 \quad \text{and} \quad \xi = \xi^+(w) > 0$$

the left and right inverses of the function $w = w(\xi)$, respectively. We have the relation $|\xi^{\pm}|'(w) = |\xi^{\pm}|'(yA_0) = \frac{1}{\kappa A_0}|\zeta^{\pm}|'(y)$ for the steepness of the inverses. Therefore, since $\kappa > 0$ and the sign of A_0 equals the sign of $-\mu$, Theorem 6.5(i) implies that $|\xi^{\pm}|'(w)$ is increasing (decreasing) in case $\mu < 0$ ($\mu > 0$) with respect to Θ_1 for any fixed κ, A_0, Θ and w.

The asymmetry of the wave at the relative level y between 0 and Θ_1 can be expressed by the ratio $\frac{|\xi^+(yA_0)|}{|\xi^-(yA_0)|}$ which, in view of the relation $\xi^{\pm}(yA_0) = \frac{1}{\kappa}\zeta^{\pm}(y)$ and Theorem 6.5 (ii), is expressed as

$$\frac{|\xi^+(yA_0)|}{|\xi^-(yA_0)|} = \bar{F}_{\Theta_1,y}(\Theta) = \bar{F}_{\Theta_1,y}\left(\frac{2(c^2-a_0)^2}{3(c^2-a_1)}\sqrt{\frac{a_0\alpha - c^2\alpha - \vartheta}{(c^2-a_0)(c^2-a_1)}}\cdot\frac{\nu}{\mu}\right).$$

The asymmetry depends on c^2, the coefficients of linear terms $a_0, a_1, \alpha, \vartheta$ and the ratio of the coefficients of nonlinear terms $\frac{\nu}{\mu}$. Note that Θ is the only canonical parameter depending on the coefficient ν. Observing (6.75) and the definition of Θ we see that Θ increases (decreases) in case I (II) if $\frac{\nu}{\mu}$ increases. Thus, by Theorem 6.5 the asymmetry $\frac{|\xi^+(yA_0)|}{|\xi^-(yA_0)|}$ is a decreasing function of the ratio $\frac{\nu}{\mu}$. By (6.76) the solitary wave in $\mathcal{W}_2$ exists for $\frac{\nu}{\mu} \in (-\Lambda, \Lambda)$ where $\Lambda = [\frac{\alpha(c^2-a_1)}{\vartheta^2(1-\frac{\alpha}{\vartheta}(a_0-c^2))}]^{3/2}$. The balance between the nonlinearity and the dispersion collapses at the critical values $\frac{\nu}{\mu} = \pm\Lambda$.

At the end of this subsection we establish the ranges for the velocity c. For this purpose we denote $q = (4\vartheta\frac{\nu^2}{\mu^2})^{1/3}$ and solve the inequalities (6.72), (6.75) and (6.76) with respect to c^2. The result is different in four subcases of the difference $a_0 - a_1$. (The first two and last two of them correspond to normal and anomalous dispersion of acoustic waves, respectively.) More precisely, the range is in case $a_0 - a_1 \geq \frac{4\vartheta}{3\alpha}$

$$c^2 \in D_q^1 \quad \text{when } 0 \leq q < 1, \qquad c^2 \in D_q^3 \quad \text{when } q \geq 1;$$

in case $\frac{\vartheta}{\alpha} < a_0 - a_1 < \frac{4\vartheta}{3\alpha}$

$$\begin{aligned} &c^2 \in D_q^1 && \text{when } 0 \leq q \leq 4 - \frac{3\alpha}{\vartheta}(a_0 - a_1),\\ &c^2 \in D_q^2 \cup D_q^1 && \text{when } 4 - \frac{3\alpha}{\vartheta}(a_0 - a_1) < q < 1,\\ &c^2 \in D_q^3 && \text{when } q \geq 1; \end{aligned}$$

in case $0 < a_0 - a_1 < \frac{\vartheta}{\alpha}$

$$\begin{aligned} &c^2 \in D_q^2 \cup D_q^1 && \text{when } 0 \leq q \leq \frac{\alpha}{\vartheta}(a_0 - a_1),\\ &c^2 \in D_q^2 && \text{when } \frac{\alpha}{\vartheta}(a_0 - a_1) \leq q < 1,\\ &c^2 \in D_q^4 && \text{when } q \geq 1; \end{aligned}$$

and in case $a_0 - a_1 \leq 0$

$$c^2 \in D_q^2 \quad \text{when } 0 \leq q < 1, \qquad c^2 \in D_q^4 \quad \text{when } q \geq 1.$$

Here

$$D_q^1 = \left(\max\left\{a_1, a_0 - \frac{\vartheta}{\alpha}, a_0 - \frac{\vartheta}{\alpha} + \frac{1}{1-q}\left[a_1 - a_0 + \frac{\vartheta}{\alpha}\right]\right\}, a_0\right),$$

$$D_q^2 = \left(a_0 - \frac{4\vartheta}{3\alpha}, \min\left\{a_1, a_0 - \frac{\vartheta}{\alpha}, a_0 - \frac{\vartheta}{\alpha} + \frac{1}{1-q}\left[a_1 - a_0 + \frac{\vartheta}{\alpha}\right]\right\}\right),$$

$$D_q^3 = \left(a_0 - \frac{\vartheta}{\alpha}, \min\left\{a_0, a_0 - \frac{\vartheta}{\alpha} + \frac{1}{q-1}\left[a_0 - a_1 - \frac{\vartheta}{\alpha}\right]\right\}\right),$$

$$D_q^4 = \left(\max\left\{a_0 - \frac{4\vartheta}{3\alpha}, a_0 - \frac{\vartheta}{\alpha} + \frac{1}{q-1}\left[a_0 - a_1 - \frac{\vartheta}{\alpha}\right]\right\}, a_0 - \frac{\vartheta}{\alpha}\right).$$

In the case of the lack of dispersion, i.e., when $a_0 - a_1 - \frac{\vartheta}{\alpha} = 0$, the range is

$$c^2 \in \left(a_0 - \frac{4\vartheta}{3\alpha}, a_1\right) \cup (a_1, a_0) \quad \text{when } 0 \le q < 1,$$

$$c^2 \in \emptyset \quad \text{when } q \ge 1.$$

6.3.4 The Case $\nu = 0$

In the symmetric case $\nu = \Theta = 0$ the equations of the solitary wave can be integrated exactly within elementary functions. Namely, the formula for the inverses of the canonical wave is in the case $\Theta_1 \in (0, 1)$

$$\begin{aligned} \zeta^{\pm}(y) = \pm\Bigg[\ln\left(\frac{2\sqrt{\Theta_1\Theta_2(\Theta_1 - y)(\Theta_2 - y)} - (\Theta_1 + \Theta_2)y + 2\Theta_1\Theta_2}{(\Theta_2 - \Theta_1)y}\right) \\ + \sqrt{\Theta_1\Theta_2}\ln\left(\frac{\Theta_1 + \Theta_2 - 2y - 2\sqrt{(\Theta_1 - y)(\Theta_2 - y)}}{\Theta_2 - \Theta_1}\right)\Bigg], \quad y \in (0, \Theta_1] \end{aligned} \tag{6.78}$$

and in the case $\Theta_1 < 0$

$$\begin{aligned} \zeta^{\pm}(y) = \pm\Bigg[\ln\left(\frac{-2\sqrt{\Theta_1\Theta_2(\Theta_1 - y)(\Theta_2 - y)} - (\Theta_1 + \Theta_2)y + 2\Theta_1\Theta_2}{(\Theta_2 - \Theta_1)y}\right) \\ - \sqrt{-\Theta_1\Theta_2}\left\{\arcsin\left(\frac{2y - \Theta_1 - \Theta_2}{\Theta_1 - \Theta_2}\right) - \frac{\pi}{2}\right\}\Bigg], \quad y \in [\Theta_1, 0). \end{aligned} \tag{6.79}$$

Examples of canonical waves for different values of Θ_1 are shown in Figs. 6.19 and 6.21. As we saw in Sect. 6.3.2, the parameter Θ_1 is related to the steepness of the wave, because $|\xi^{\pm}|'(y)$ is increasing in Θ_1. Although the value of Θ_1 is also the amplitude of $y(\zeta)$, the wave is not proportional to Θ_1. Numerical results suggest

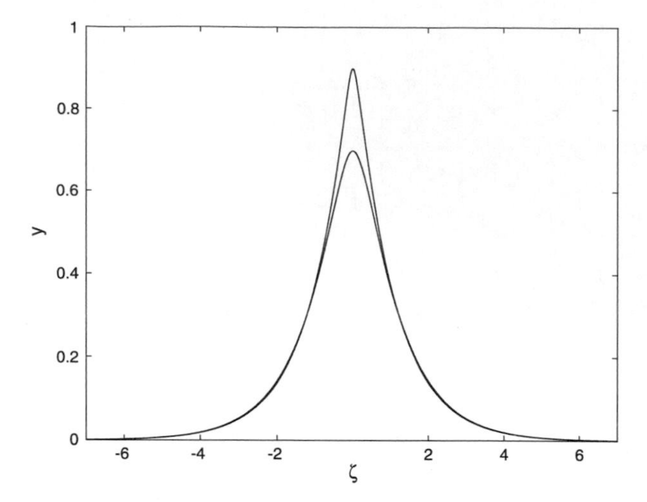

Fig. 6.19 Canonical wave $y(\zeta)$ in cases $\Theta_1 = 0.9$ and $\Theta_1 = 0.7$

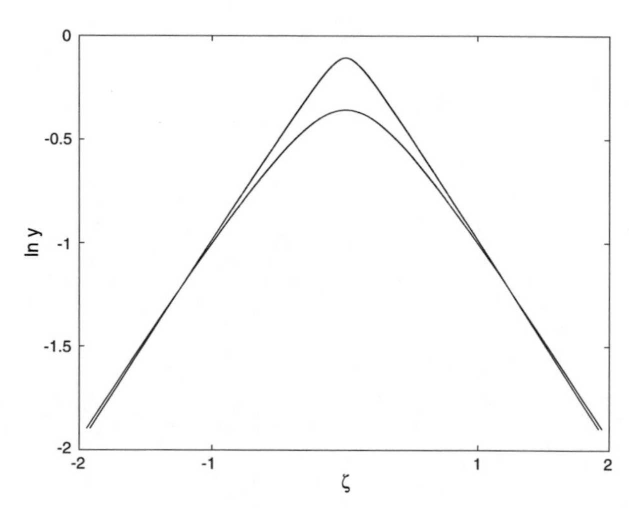

Fig. 6.20 Function $\ln y(\zeta)$ in cases $\Theta_1 = 0.9$ and $\Theta_1 = 0.7$

that all waves corresponding to $\Theta_1 \in (0, 1)$ intersect with each other. This is well seen from graphs in logarithmic scale (e.g. Fig. 6.20).

Finally, for the inverses of the general wave we have the formulas in case I

$$\xi^{\pm}(w) = \pm\frac{1}{\kappa}\Bigg[\ln\left(\frac{2(c^2-a_0)\sqrt{1-\frac{\mu(1-p)}{c^2-a_0}w+\frac{\mu^2(1-3p)}{4(c^2-a_0)^2}w^2}-\mu(1-p)w+2(c^2-a_0)}{\mu\sqrt{p+p^2}w}\right) + \frac{2}{\sqrt{1-3p}}\ln\left(\frac{1-p-\frac{\mu(1-3p)}{2(c^2-a_0)}w-\sqrt{1-3p}\sqrt{1-\frac{\mu(1-p)}{c^2-a_0}w+\frac{\mu^2(1-3p)}{4(c^2-a_0)^2}w^2}}{\sqrt{p+p^2}}\right)\Bigg],$$

$$A^{-1}w \in (0, 1], \tag{6.80}$$

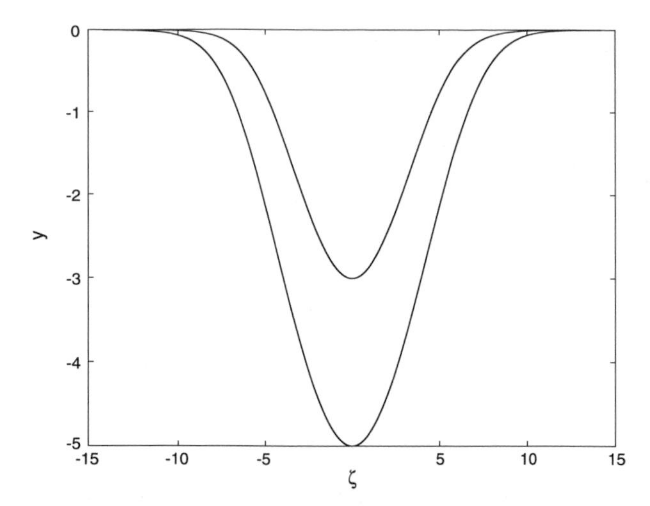

Fig. 6.21 Canonical wave $y(\zeta)$ in cases $\Theta_1 = -5$ and $\Theta_1 = -3$

and in case II

$$\xi^{\pm}(w)$$

$$= \pm\frac{1}{\kappa}\left[\ln\left(\frac{2(c^2-a_0)\sqrt{1-\frac{\mu(1-p)}{c^2-a_0}w+\frac{\mu^2(1-3p)}{4(c^2-a_0)^2}w^2}+\mu(1-p)w-2(c^2-a_0)}{\mu\sqrt{p+p^2}w}\right)\right.$$

$$\left.-\frac{2}{\sqrt{3p-1}}\left\{\arcsin\left(\frac{\frac{\mu(1-3p)}{2(c^2-a_0)}w-1+p}{\sqrt{p+p^2}}\right)-\frac{\pi}{2}\right\}\right],$$

$$A^{-1}w \in (0,1], \tag{6.81}$$

where κ and p are given in terms of $c^2, a_0, a_1, \vartheta, \alpha$ by the formulas (6.49) and (6.50).

In case $\Theta \neq 0$ the solution of (6.58) is a higher transcendental function. It is possible to expand it as a Taylor series with respect to Θ as in the case of the hierarchical equation in Sect. 6.2.4. But this series is very complicated and hard to use practically. Therefore, we will not deduce it here.

6.3.5 Comparison with Hierarchical Equation

The coefficients of the hierarchical equation (6.3) are given in terms of the coefficients of the coupled system (6.40) by the formulas (3.45). This connection enables us to compare the existence conditions for the solitary waves in these models that are given in Theorems 6.3 and 6.6, respectively. It is easy to check that the conditions for the nonlinearity ratio, i.e., (6.28) and (6.76), coincide in these theorems. However, the conditions for the velocity c are different. In Theorem 6.3 one assumes the simple non-vanishing conditions $\mu \neq 0$, $\beta c^2 - \gamma \neq 0$ and $c^2 - b \neq 0$.

But in Theorem 6.6 the stronger inequalities (6.72) and (6.75) are to be satisfied. Therefore, the assumptions of Theorem 6.6 imply the assumptions of Theorem 6.3 but not vice versa. The existence conditions for solitary wave solutions are in the case of the coupled system stronger than in the case of the hierarchical equation.

6.4 Proofs of Mathematical Statements

6.4.1 Proofs of Sect. 6.2

Proof of Lemma 6.1 Let (6.5) have a solution $w \in \mathcal{W}_4$. First we prove the relation $\beta c^2 - \gamma \neq 0$. Suppose on the contrary that $\beta c^2 - \gamma = 0$. Then (6.8) has the form

$$\delta^{3/2}\lambda w' w'' = \left(c^2 - b - \frac{\mu}{2} w\right) w. \tag{6.82}$$

Evidently, in this case $\lambda \neq 0$ because otherwise the left-hand side of (6.8) is zero which implies that w is a constant and cannot belong to $\mathcal{W}_4$. Since $w(\xi) \to 0$ as $|\xi| \to \infty$, there exists a point $\xi_1 \in \mathbb{R}$ where the function $|w|$ attains the absolute maximum. It holds $w'(\xi_1) = 0$. This by (6.82) gives $w(\xi_1) = w_0 := 2(c^2 - b)/\mu$. Further, observing that $w'(\xi) \to 0$ as $|\xi| \to \infty$, we see that $|w'|$ has an absolute maximum at a point $\xi_2 \in \mathbb{R}$. We get $w''(\xi_2) = 0$ and by (6.82) either $w(\xi_2) = w_0$ or $w(\xi_2) = 0$ holds. In the former case the function $|w|$ attains values greater than $|w_0|$ in a neighbourhood of ξ_2, because $w'(\xi_2) \neq 0$. This contradicts to the proved statement that $|w_0|$ is the absolute maximum of $|w|$. In the latter case the function w changes sign at ξ_2. But then w has positive and negative values on the line $\mathbb{R}$, and in view of the relation $w(\xi) \to 0$ as $|\xi| \to \infty$ there exist $\xi_3, \xi_4 \in \mathbb{R}$ such that $w(\xi_3) < 0$, $w(\xi_4) > 0$ and $w'(\xi_3) = w'(\xi_4) = 0$. Now the relation (6.82) implies that $w(\xi_j) \in \{0; w_0\}$, $j = 3, 4$. But this contradicts the inequalities $w(\xi_3) < 0$ and $w(\xi_4) > 0$. Summing up, the supposition $\beta c^2 - \gamma = 0$ was wrong. The inequality $\beta c^2 - \gamma \neq 0$ is valid.

Next, let us prove that w' and $\delta(\beta c^2 - \gamma) + \delta^{3/2}\lambda w'$ may vanish only in a countable subset of $\mathbb{R}$. Suppose on the contrary that there exists an interval (ξ_1, ξ_2), such that

$$\begin{array}{lll} \text{either} & \text{(a) } w'(\xi) = 0 & \text{for all } \xi \in (\xi_1, \xi_2) \\ \text{or} & \text{(b) } \delta(\beta c^2 - \gamma) + \delta^{3/2}\lambda w'(\xi) = 0 & \text{for all } \xi \in (\xi_1, \xi_2). \end{array} \tag{6.83}$$

In case (a) the function w is constant on the interval $[\xi_1, \xi_2]$, and from (6.8) in view of $\beta c^2 - \gamma \neq 0$, we see that w satisfies the ordinary differential equation $w'' = f(w, w')$ with smooth right-hand side $f(w, w') = \frac{(c^2-b)w - \frac{\mu}{2}w^2}{\delta(\beta c^2-\gamma)+\delta^{3/2}\lambda w'}$ in neighbourhoods of the points $\xi = \xi_1$ and $\xi = \xi_2$. By Cauchy theorem we can extend the solution w of this equation uniquely as a constant to the whole line $\mathbb{R}$. But this is in contradiction with the definition of the space $\mathcal{W}_4$ that does not contain constant functions. In case (b) we firstly note that $\lambda \neq 0$, because of $\delta(\beta c^2 - \gamma) \neq 0$. This implies that $w(\xi)$ is a linear function with a nonzero slope $-(\beta c^2 - \gamma)\delta^{-1/2}\lambda^{-1}$ for any $\xi \in (\xi_1, \xi_2)$. On the other hand, using (6.83) in (6.8) we obtain $(c^2 - b)w - \frac{\mu}{2}w^2 = 0$ for $\xi \in (\xi_1, \xi_2)$. This implies that $w(\xi)$ is constant for $\xi \in (\xi_1, \xi_2)$. We have reached a contradiction. Summing up, the supposition (6.83) was wrong, and the lemma is proved. □

Proof of Lemma 6.2 As in the proof of Lemma 6.1, let $\xi_1 \in \mathbb{R}$ be a point where $|w|$ attains its absolute maximum. Then $w'(\xi_1) = 0$. If we suppose that either $c^2 - b = 0$, $\mu \neq 0$ or $c^2 - b \neq 0$, $\mu = 0$ then (6.9) implies that $w(\xi_1) = 0$. But this equality cannot hold at the absolute maximum of $|w|$ due to the non-triviality of $w \in \mathcal{W}_4$. If we suppose that $c^2 - b = \mu = 0$ then (6.9) gives $w' \equiv \text{const}$, which is also not the case for $w \in \mathcal{W}_4$. Thus, $c^2 - b \neq 0$ and $\mu \neq 0$.

Noting that $w, w' \to 0$ as $|\xi| \to \infty$ we see that the asymptotic relation

$$\left(w'\right)^2 \sim \frac{c^2 - b}{\delta(\beta c^2 - \gamma)} w^2 \quad \text{as } |\xi| \to \infty$$

is valid for the solution. This implies $\frac{c^2 - b}{\beta c^2 - \gamma} > 0$. The proof is complete. □

Proof of Theorem 6.2 The differentiation of the equation of the trajectory $z^2 - \Theta z^3 = y^2 - y^3$ with respect to Θ yields

$$2z \frac{dz}{d\Theta} - 3\Theta z^2 \frac{dz}{d\Theta} - z^3 = 0.$$

Solving this equation for $\frac{dz}{d\Theta}$ we obtain

$$\frac{dz}{d\Theta} = \frac{z^2}{2 - 3\Theta z}. \tag{6.84}$$

Since $z < \frac{2}{3\Theta}$ on the trajectory T (Fig. 6.6), the inequality $\frac{dz}{d\Theta} > 0$ holds for any $z \neq 0$. Noting that $z = y'$, we see that the derivative $y'(\xi)$ increases in Θ for any $\xi \neq 0$, because $\xi = 0$ is the single stationary point of the solution $y(\xi)$ (Fig. 6.7). This implies that derivatives $\zeta^{-'}(y)$ and $\zeta^{+'}(y)$ of the inverses of the solution are decreasing in Θ for any $y \neq 1$. Observing in addition the signs of these inverses we reach the following relations:

$$\begin{aligned} &\zeta^{-'}(y) \quad \text{is positive and decreasing in } \Theta \\ -&\zeta^{+'}(y) \text{ is positive and increasing in } \Theta. \end{aligned} \tag{6.85}$$

Thus, denoting the asymmetry at the level $y \in (0, 1)$ by F_y, we can express it as follows:

$$F_y(\Theta) := \frac{|\zeta^+(y)|}{|\zeta^-(y)|} = \left| \frac{\int_1^y \zeta^{+'}(s)\, ds}{\int_1^y \zeta^{-'}(s)\, ds} \right| = \frac{\int_y^1 [-\zeta^{+'}(s)]\, ds}{\int_y^1 \zeta^{-'}(s)\, ds}. \tag{6.86}$$

Due to (6.85), $F_y(\Theta)$ is increasing. In the case $\Theta = 0$ the solution is symmetric (see (6.19)). Therefore, $F_y(0) = 1$ and the theorem is proved. □

6.4.2 Proofs of Sect. 6.3

Proof of Lemma 6.3 First of all, we remark that at least one of the quantities $c^2 - a_0$ or μ is different from zero. Indeed, otherwise the left-hand side of (6.44) is zero for any ξ and hence the function w is constant and cannot belong to $\mathcal{W}_2$. Consequently, for any $w \in \mathcal{W}_2$ the function

$$\hat{w} = \left(c^2 - a_0\right)w - \frac{\mu}{2}w^2 \tag{6.87}$$

also belongs to $\mathcal{W}_2$.

Let prove the assertion $c^2 - a_1 \neq 0$. Suppose that $c^2 - a_1 = 0$. In this case (6.44) becomes

$$\delta^{3/2}\nu\hat{w}'\hat{w}'' = \alpha\left\{\left(c^2 - a_0\right)w(\xi) - \frac{\mu}{2}w^2(\xi)\right\} + \vartheta w(\xi). \tag{6.88}$$

Note that $\nu \neq 0$. Indeed, if $\nu = 0$ then w is a constant, again. Since $\hat{w} \in \mathcal{W}_2$, the function $|\hat{w}|$ has an absolute maximum at some point $\xi_1 \in \mathbb{R}$. This yields $\hat{w}'(\xi_1) = 0$ and from (6.88) we have $w(\xi_1) = w_1 := 2(c^2\alpha - a_0\alpha + \vartheta)/\mu$. Thus, $\hat{w}(\xi_1) = \hat{w}_1 := (c^2 - a_0)w_1 - \frac{\mu}{2}w_1^2$. Moreover, $|\hat{w}'|$ has an absolute maximum at a point $\xi_2 \in \mathbb{R}$. This means that $\hat{w}''(\xi_2) = 0$ and (6.88) implies that $\hat{w}(\xi_2) \in \{\hat{w}_1; 0\}$. In the case $\hat{w}(\xi_2) = \hat{w}_1$, the function $|w|$ attains values greater than $|\hat{w}_1|$ in a neighbourhood of ξ_2. This contradicts the fact that $|\hat{w}_1|$ is the absolute maximum of $|\hat{w}|$. In the case $\hat{w}(\xi_2) = 0$, the function $\hat{w}$ changes sign at ξ_2. Then there exist $\xi_3, \xi_4 \in \mathbb{R}$, such that $\hat{w}(\xi_3) < 0$, $\hat{w}(\xi_4) > 0$ and $\hat{w}'(\xi_3) = \hat{w}'(\xi_4) = 0$. Due to (6.88) $w(\xi_j) \in \{0; w_1\}$, $j = 3, 4$, and hence $\hat{w}(\xi_j) \in \{0; \hat{w}_1\}$, $j = 3, 4$. But this cannot hold because the quantities $\hat{w}(\xi_j)$, $j = 3, 4$ have different signs. Summing up, the supposition $c^2 - a_1 = 0$ was wrong. We have $c^2 - a_1 \neq 0$.

Before we continue with the proof, let us note that any function $w \in \mathcal{W}_2$ solving (6.44) solves (6.46), too. Equation (6.46) is obtained from (6.44) by multiplication by $[(c^2 - a_0)w - \frac{\mu}{2}w^2]'$ and integration. (So far, we have the one-sided implication (6.44) $\Rightarrow$ (6.46), only!).

Next let us prove that $\mu \neq 0$. If $\mu = 0$ then $\frac{\alpha}{2}(c^2 - a_0) + \vartheta \neq 0$ because otherwise the right-hand side of (6.46) is identically zero and either w or w' is a constant function. Further, the function w in $\mathcal{W}_2$ must have at least one argument ξ_0 where $w'(\xi_0) = 0$ and $w(\xi_0) \neq 0$. But then in view of $\mu = 0$ and $\frac{\alpha}{2}(c^2 - a_0) + \vartheta \neq 0$ the left-hand side of (6.46) is zero but the right-hand side is different from zero at $\xi = \xi_0$. This is the contradiction. Consequently, $\mu \neq 0$.

Further, let us prove that $c^2 - a_0 \neq 0$. Suppose that $c^2 - a_0 = 0$. Then (6.46) can be written

$$\frac{\delta(c^2 - a_1)}{2}\mu^2 w^2\left[w'\right]^2 + \frac{\delta^{3/2}\nu}{3}\mu^3 w^3\left[w'\right]^3 + \frac{\alpha}{2}\frac{\mu^2}{4}w^4 - \frac{\vartheta\mu}{3}w^3 = 0.$$

Since w is continuous, we can define the following quantity:

$$s = \begin{cases} \infty & \text{if } w(\xi) \neq 0 \text{ a.e.} \\ d & \text{otherwise} \end{cases}$$

where d is some number such that $w(\xi) \neq 0$ a.e. in $(d-\delta, d)$ and $w(\xi) = 0$ in $(d, d+\delta)$ with some $\delta > 0$. Now we can choose an increasing sequence $\tilde{\xi}_i \to s^-$ such that $w(\tilde{\xi}_i) \neq 0$, $w'(\tilde{\xi}_i) \neq 0$ and $w(\tilde{\xi}_n)$, $w'(\tilde{\xi}_n)$, $w''(\tilde{\xi}_n) \to 0$. In the case of finite s the existence of such a sequence easily follows from the continuity of w, w', w'' and in the case $s = \infty$ this follows from the relations $w, w' \to 0$ as $\xi \to \infty$ and $w(\xi) \neq 0$ a.e. By the continuity, $w(\xi) \neq 0$ in neighbourhoods of the points $\tilde{\xi}_i$, as well. Thus, we can divide (6.88) by w^2 in these neighbourhoods to get

$$\frac{\delta(c^2-a_1)}{2}\mu^2\left[w'\right]^2 + \frac{\delta^{3/2}\nu}{3}\mu^3 w\left[w'\right]^3 + \frac{\alpha}{2}\frac{\mu^2}{4}w^2 - \frac{\vartheta\mu}{3}w = 0.$$

We differentiate this equation in these neighbourhoods, set $\xi = \tilde{\xi}_i$ and divide by $w'(\tilde{\xi}_i)$. This results in the following relation:

$$\left.\left\{\delta(c^2-a_1)\mu^2 w'' + \frac{\delta^{3/2}\nu}{3}\mu^3\left[ww'w'' + \left[w'\right]^3\right] + \frac{\alpha\mu^2}{4}w - \frac{\vartheta\mu}{3}\right\}\right|_{\xi=\tilde{\xi}_i} = 0.$$

Passing to the limit $\tilde{\xi}_i \to s^-$ we obtain $-\frac{\vartheta\mu}{3} = 0$. But this cannot be valid, because $\vartheta \neq 0$ and $\mu \neq 0$. Thus, we have proved that $c^2 - a_0 \neq 0$.

Next we prove the additional inequality

$$\frac{\alpha}{2}\left(c^2 - a_0\right) + \vartheta \neq 0. \tag{6.89}$$

Setting $\frac{\alpha}{2}(c^2 - a_0) + \vartheta = 0$, the equation (6.44) has the form

$$\left[\delta\left(c^2-a_1\right) - \delta^{3/2}\nu\hat{w}'\right]\hat{w}'' = -\frac{\alpha}{2}\left\{\left(c^2-a_0\right)w - \mu w^2\right\}.$$

Multiplying by w' we get

$$\left[\delta\left(c^2-a_1\right) - \delta^{3/2}\nu\hat{w}'\right]w'\hat{w}'' = -\frac{\alpha}{2}w\hat{w}'. \tag{6.90}$$

Since $\hat{w} \in \mathcal{W}_2$, there exists $\hat{\xi}_1$ such that $\hat{w}''(\hat{\xi}_1) = 0$ and $\hat{w}'(\hat{\xi}_1) \neq 0$. From (6.90) we see that $w(\hat{\xi}_1) = 0$. This by (6.87) implies that $\hat{w}(\hat{\xi}_1) = 0$. Since $\hat{w}'(\hat{\xi}_1) \neq 0$ the function $\hat{w}$ is not identically zero for $\xi > \hat{\xi}_1$. Taking the relations $\hat{w}(\hat{\xi}_1) = 0$ and $\lim_{\xi\to\infty}\hat{w}(\xi) = 0$ into account, we see that there exists a next point $\hat{\xi}_2 > \hat{\xi}_1$ such that $\hat{w}''(\hat{\xi}_2) = 0$ and $\hat{w}'(\xi_2) \neq 0$. Again, (6.90) yields $w(\hat{\xi}_2) = 0$ and from (6.87) we have $\hat{w}(\hat{\xi}_2) = 0$. Continuing this process we obtain a sequence of numbers $\hat{\xi}_1 < \hat{\xi}_2 < \hat{\xi}_3 < \cdots$ such that

$$\hat{w}(\hat{\xi}_j)' \neq 0, \qquad \hat{w}(\hat{\xi}_j) = w(\hat{\xi}_j) = 0, \quad j = 1, 2, \ldots.$$

Since $\hat{w}$ is not identically zero on intervals $(\hat{\xi}_j, \hat{\xi}_{j+1})$, $j = 1, 2, \ldots$, the function w is also not identically zero on these intervals. Since $w(\hat{\xi}_j) = 0$ we see that there exist $\bar{\xi}_j \in (\hat{\xi}_j, \hat{\xi}_{j+1})$, $j = 1, 2, \ldots$, such that $w'(\bar{\xi}_j) = 0$ and $w(\bar{\xi}_j) \neq 0$. Further, we note that the limit $\lim w(\bar{\xi}_j) = 0$ is valid. In the case of an unbounded sequence $\bar{\xi}_j$ this follows from the condition $\lim_{\xi \to \infty} w(\xi) = 0$ in the definition of $\mathcal{W}_2$, and in the case of bounded sequence $\bar{\xi}_j$, this follows from the boundedness of the derivative of w in $\mathcal{W}_2$ and the relation $|\bar{\xi}_{j+1} - \bar{\xi}_j| \to 0$. We point out that the vanishing sequence $w(\bar{\xi}_j) \neq 0$ must contain infinitely many different numbers. On the other hand, let us consider the equation (6.46). Since $w'(\bar{\xi}_j) = 0$, the left-hand side of (6.46) is zero at $\xi = \bar{\xi}_j$. Thus, $\mathcal{P}w(\bar{\xi}_j) = 0$ where $\mathcal{P}w$ stands for the right-hand side of (6.46). But $\mathcal{P}w$ is a non-trivial 4-th order polynomial of w which may have maximally 4 different roots. This implies that the sequence $w(\bar{\xi}_j)$ may contain maximally 4 different numbers. We have reached the contradiction. This proves the desired inequality (6.89).

Let us prove the remaining assertions of the lemma. Suppose that any of them does not hold. This means that there exists an interval (ξ_1, ξ_2) such that

$$\begin{aligned} &\text{either} \quad (a)\ c^2 - a_0 - \mu w(\xi) = 0 \quad \text{for all } \xi \in (\xi_1, \xi_2) \\ &\text{or} \quad (b)\ \hat{w}'(\xi) = 0 \quad \text{for all } \xi \in (\xi_1, \xi_2) \\ &\text{or} \quad (c)\ \delta(c^2 - a_1) - \delta^{3/2} \nu \hat{w}'(\xi) = 0 \quad \text{for all } \xi \in (\xi_1, \xi_2). \end{aligned} \tag{6.91}$$

Note that (a) implies (b). Therefore, it is sufficient to deal with the cases (b) and (c) only. From (6.87) we obtain the range $2\mu\hat{w} \leq (c^2 - a_0)^2$ for the variable $\hat{w}$. Under this condition the equation (6.87) has a unique w solution vanishing at infinity:

$$w = z(\hat{w}) \quad \text{where } z(\hat{w}) = \frac{c^2 - a_0}{\mu}\left[1 - \sqrt{1 - \frac{2\mu\hat{w}}{(c^2 - a_0)^2}}\,\right].$$

Let (b) hold. Then the function g is continuously differentiable in a neighbourhood of the points ξ_1 and ξ_2. Indeed, g' may have a discontinuity only at $\hat{w} = \frac{(c^2 - a_0)^2}{2\mu}$ when $w = g(w) = \frac{c^2 - a_0}{\mu}$. But this is not the case at ξ_1 and ξ_2 because there $\hat{w}'' = 0$ which by (6.44) gives $\alpha\hat{w} + \vartheta w = 0$. But due to the proved inequality (6.89) we have $\alpha \frac{(c^2 - a_0)^2}{2\mu} + \vartheta \frac{c^2 - a_0}{\mu} \neq 0$. From (6.44) we deduce the equation $\hat{w}'' = f(\hat{w}', \hat{w})$ for $\hat{w}$ where

$$f(\hat{w}', \hat{w}) = -\frac{\alpha\hat{w} + \vartheta g(\hat{w})}{\delta(c^2 - a_1) - \delta^{3/2}\nu\hat{w}'}.$$

In case (b) the function f is continuously differentiable in neighbourhoods of the points ξ_1 and ξ_2. This follows from the proved inequalities $c^2 - a_1 \neq 0$ and the regularity of $g(w)$. By Cauchy theorem we can extend the solution w of this equation uniquely as a constant to the whole line $\mathbb{R}$. This contradicts the definition of the space $\mathcal{W}_2$ that does not contain constant functions. Thus, (b) is wrong. In case (c) we firstly note that $\nu \neq 0$, because of $c^2 - a_1 \neq 0$. This yields that $\hat{w}(\xi)$ is a linear function with a nonzero slope in (ξ_1,ξ_2). On the other hand, using (6.91) in (6.44) we obtain $\alpha\{(c^2-a_0)w-\frac{\mu}{2}w^2\}+\vartheta w=0$ for $\xi\in(\xi_1,\xi_2)$. This implies that $w(\xi)$ is constant in (ξ_1,ξ_2) and hence by (6.87) $\hat{w}(\xi)$ is constant in (ξ_1,ξ_2). We have reached a contradiction. Thus, (c) is wrong. Lemma is completely proved. □

Proof of Lemma 6.4 Since $w,w'\to 0$ as $|\xi|\to\infty$, from (6.47) with (6.48) we deduce the asymptotic relation

$$(w')^2\sim\frac{a_0\alpha-c^2\alpha-\vartheta}{\delta(c^2-a_1)(c^2-a_0)}w^2\quad\text{as } |\xi|\to\infty.$$

This implies that $\frac{a_0\alpha-c^2\alpha-\vartheta}{\delta(c^2-a_1)(c^2-a_0)}\geq 0$. To prove (6.49), we have to show that in addition $a_0\alpha-c^2\alpha-\vartheta\neq 0$. Supposing $a_0\alpha-c^2\alpha-\vartheta=0$ we substitute ϑ by $\alpha(a_0-c^2)$ in the formula (6.48) and transform it to the form

$$P(w)=\frac{\mu\alpha w^3}{3\delta(c^2-a_1)(c^2-a_0)}\left[1-\frac{3\mu}{4(c^2-a_0)}w\right].$$

According to the definition of $\mathcal{W}_2$, the function $|w|\in\mathcal{W}_2$ must have a maximum point where $w'=0$ but $w\neq 0$. At such a point the left-hand side of (6.47) is zero. Hence due to the formula of $P(w)$ we have $w=\frac{4(c^2-a_0)}{3\mu}$ there. Further, since the function w is continuous and approaches 0 as $\xi\to\infty$, it must attain the value $\frac{c^2-a_0}{\mu}$, which is located between $\frac{4(c^2-a_0)}{3\mu}$ and 0. But for $w=\frac{c^2-a_0}{\mu}$ the left-hand of (6.47) is zero and the right-hand side not zero. This is the contradiction. We have proved (6.49).

Next, let us prove (6.50). In the case $p+p^2<0$, the quadratic function inside the square brackets in (6.48) has no real roots. Then, from (6.47) and (6.48) we see that $w'=0$ implies $w=0$. But this is in contradiction with the properties of $w\in\mathcal{W}_2$. Thus, $p+p^2\geq 0$ which implies that $p\in(-\infty,-1]\cup[0,\infty)$. If $p=-1$ then the polynomial P reads $P(w)=\kappa^2w^2(1-\frac{\mu}{c^2-a_0}w)^2$. Since $1-\frac{\mu}{c^2-a_0}w\neq 0$ a.e. (see Lemma 6.3), we can divide both sides of (6.47) by $(1-\frac{\mu}{c^2-a_0})^2$. This leads to an equation that yields the contradictive implication $w'=0\Rightarrow w=0$, again. Thus, $p\neq -1$. If $p=\frac{1}{3}$ then from the definition of p in (6.50) we immediately obtain $\alpha(c^2-a_0)=0$. But this cannot be valid because $\alpha\neq 0$ and $c^2-a_0\neq 0$ by Lemma 6.3. Therefore, $p\neq\frac{1}{3}$. In order to finish the proof, it remains to show that $p\notin[0,\frac{1}{3})$. Suppose on the contrary that $p\in[0,\frac{1}{3})$. Then the polynomial $P(w)$ has the double root 0 and two real roots $w_\pm=\frac{c^2-a_0}{\mu}\frac{2}{1-p\pm\sqrt{p+p^2}}$ such that $\frac{2}{1-p\pm\sqrt{p+p^2}}>1$. Therefore, at the maximum point of the function $|w|$, where $w'=0$, either $w=w_+$ or $w=w_-$. Again, by the continuity and the limit $\lim_{\xi\to\infty}w(\xi)$ the function w must attain the value $\frac{(c^2-a_0)}{\mu}$, which is located between the attained $w\in\{w_+;w_-\}$ and 0. For $w=\frac{(c^2-a_0)}{\mu}$ the left-hand side of (6.47) equals zero but the right-hand side is not zero. Again, we have the contradiction. Thus, $p \notin [0, \frac{1}{3})$. The lemma is proved. □

Proof of Lemma 6.5 Let $y \in \mathcal{W}_2$ solve (6.61). Then Lemma 6.3 implies that $1 - y = 1 - \frac{\mu}{c^2 - a_0} w \neq 0$ a.e. Let us show that the inequality $1 - y(\zeta) \neq 0$ holds even for all $\zeta \in \mathbb{R}$. To this end, suppose that $1 - y(\zeta_0) = 0$ for some ζ_0. Then the left-hand side of (6.61) is zero at $\zeta = \zeta_0$. But the right-hand side of (6.61) cannot be zero at this value of ζ because $y = 1$ is not a root of the polynomial $y^2(1 - \frac{y}{\Theta_1})(1 - \frac{y}{\Theta_2})$. Indeed, $\Theta_1 \neq 1$ and $\Theta_2 \neq 1$ by (6.56). Therefore, $1 - y \neq 0$ everywhere. This in view of the relation $y \to 0$ as $\zeta \to \infty$ and the continuity of y implies that

$$1 - y(\zeta) \in [\epsilon_1, \epsilon_2] \quad \text{for all } \zeta \in \mathbb{R} \text{ with some } \epsilon_2 > \epsilon_1 > 0.$$

Thus, from (6.62) we obtain $\eta'(\zeta) \in [M_1, M_2]$ with $M_i = \frac{1}{\epsilon_i}$. Clearly, $y \in C^2(\mathbb{R})$ yields $\hat{y} \in C^2(\mathbb{R})$. Moreover, $\eta \to \pm\infty$ implies that $y \to \pm\infty$. Thus, we have the convergence relations $\hat{y}, \hat{y}' \to 0$ as $|\eta| \to \infty$, too. This proves that $\hat{y} \in \mathcal{W}_2$. The differential equation in (6.64) can be deduced from (6.61) by a simple substitution, and the inequality $1 - \Theta_1 \hat{y} > 0$ immediately follows from $1 - y(\zeta) > 0$ and $\hat{y}(\eta) = \frac{1}{\theta_1} y(\zeta)$.

Conversely, let $\hat{y} \in \mathcal{W}_2$ solve (6.64). Then, the relation $1 - \Theta_1 \hat{y} > 0$ with $\lim_{\eta \to \infty} \hat{y}(\eta) = 0$ implies that

$$1 - \Theta_1 \hat{y}(\zeta) \in [\hat{M}_1, \hat{M}_2] \quad \text{for all } \eta \in \mathbb{R} \text{ with some } \hat{M}_2 > \hat{M}_1 > 0.$$

Therefore, from (6.63) we have $\zeta'(\eta) \in [\hat{M}_1, \hat{M}_2]$ and by the same arguments as above we prove $y \in \mathcal{W}_2$ and deduce (6.61) for y. □

Proof of Theorem 6.5 According to (6.55) the equation for the functions $\zeta = \zeta^{\pm}$ can be written

$$\left\{(1 - y)(\zeta')^{-1}\right\}^2 - \Theta\left\{(1 - y)(\zeta')^{-1}\right\}^3 = y^2\left(1 - \frac{y}{\Theta_1}\right)\left(1 - \frac{y}{\Theta_2}\right). \tag{6.92}$$

Let us fix some $\Theta_1^{-1} y \in (0, 1)$ and differentiate this equation with respect to Θ_1. Observing the dependence of Θ_2 on Θ_1 (see (6.55)) we obtain

$$-\left\{2 - 3\Theta(1 - y)(\zeta')^{-1}\right\}(1 - y)^2(\zeta')^{-3} \frac{d}{d\Theta_1}\zeta' = 2y^3(4 - 3y)\frac{(1 - \Theta_1)(2 - \Theta_1)}{\Theta_1^2(3\Theta_1 - 4)^2}.$$

This in view of $|\zeta^{\pm}|' = \pm\zeta^{\pm'}$ yields

$$\frac{d}{d\Theta_1}|\zeta^{\pm}|' = \frac{2y^2(4 - 3y)(\zeta^{\pm'})^2(1 - \Theta_1)(2 - \Theta_1)}{\{2 - 3\Theta(1 - y)(\zeta^{\pm'})^{-1}\}(1 - y)^2\Theta_1^2(3\Theta_1 - 4)^2} \cdot \left\{-y|\zeta^{\pm}|'\right\}.$$

Since $y < 1$, $\Theta_1 < 1$ and $2 - 3\Theta(1 - y)(\zeta')^{-1} = 2 - 3\Theta(1 - y)y' > 0$, the first factor on the right-hand side of this relation is positive. Moreover, due to the properties of y

(cf. Theorem 6.4), the second factor $-y|\zeta^{\pm}|'$ is also positive. Thus, we see that $|\zeta^{\pm}|'$ is increasing with respect to Θ_1. This proves (i).

To prove (ii), let us differentiate (6.92) with respect to Θ and solve it for $\frac{d}{d\Theta}\zeta'$:

$$\frac{d}{d\Theta}\zeta' = -\frac{1-y}{2-3\Theta(1-y)(\zeta')^{-1}}.$$

Since the right-hand side of this expression is negative, the functions $\zeta^{-'}(y)$ and $\zeta^{+'}(y)$ are decreasing in Θ. Observing in addition the signs of these inverses we see that (6.85) holds when $0 < \Theta_1 < 1$ and, when $\Theta_1 < 0$,

$$\begin{aligned} &-\zeta^{-'}(y) \text{ is positive and increases in } \Theta, \\ &\zeta^{+'}(y) \text{ is positive and decreases in } \Theta. \end{aligned} \tag{6.93}$$

Further, for any y between 0 and Θ_1 we have the formula

$$\begin{aligned} F_{\Theta_1,y}(\Theta) &:= \frac{|\zeta^+(y)|}{|\zeta^-(y)|} \\ &= \left|\frac{\int_{\Theta_1}^{y}\zeta^{+'}(s)\,ds}{\int_{\Theta_1}^{y}\zeta^{-'}(s)\,ds}\right| = \begin{cases} \dfrac{\int_y^{\Theta_1}[-\zeta^{+'}(s)]ds}{\int_y^{\Theta_1}\zeta^{-'}(s)\,ds} & \text{when } 0 < \Theta_1 < 1 \\ \dfrac{\int_{\Theta_1}^{y}\zeta^{+'}(s)\,ds}{\int_{\Theta_1}^{y}[-\zeta^{-'}(s)]ds} & \text{when } \Theta_1 < 0. \end{cases} \end{aligned}$$

In view of (6.85) and (6.93) the function $F_{\Theta_1,y}$ is increasing (decreasing) when $0 < \Theta_1 < 1$ ($\Theta_1 < 0$) for $\Theta \in (-b_{\Theta_1}, b_{\Theta_1})$. Finally, $\zeta^+ = -\zeta^-$ when $\Theta = 0$ and hence $F_y(0) = 1$. The theorem is proved. □

Chapter 7
Inverse Problems for Solitary Waves

7.1 Inverse Problems for Hierarchical Equation

7.1.1 Formulation of Inverse Problems

We are going to investigate the possibilities of determining the five coefficients b, μ, β, γ and λ of the hierarchical equation (3.36) by means of measurements gathered from solitary waves.

We remark that a single solitary wave does not contain enough information to reconstruct all five unknowns. The reason is that the solution of (6.11) of such a wave has only three degrees of freedom (the parameters A, κ and Θ). Even if we have measured the whole wave $w(\xi)$, we can expect to recover only those three parameters A, κ and Θ. The system (6.10) has infinitely many solutions $b, \mu, \beta, \gamma, \lambda$ for given A, κ, Θ and c^2. Consequently, it is necessary to measure at least two waves with different values of c^2. In the sequel we will show that this is enough to recover all unknowns.

Let $w[c_1]$ and $w[c_2]$ be two solitary waves with the velocities c_1 and c_2 satisfying $c_1^2 \neq c_2^2$. Suppose that we know the amplitudes of these waves. Denote them by A_1 and A_2, respectively. Then, from (6.10) we deduce the system

$$3b + A_j\mu = 3c_j^2, \quad j = 1, 2, \tag{7.1}$$

for the unknowns b and μ. The assumption $c_1^2 \neq c_2^2$ yields $A_1 \neq A_2$. Therefore, the system (7.1) is regular. This implies that the coefficients b and μ are uniquely recovered by amplitudes of two waves. We see that the determination of the pair b, μ is rather a trivial task.

Now let the quantities b and μ be known and ask the question: how to reconstruct the remaining parameters β, γ and λ? Here we can distinguish between two cases: (1) the waves are symmetric—then $\lambda = 0$ and the number of unknowns reduces to two; (2) the waves are asymmetric—then $\lambda \neq 0$ and we have to find the full triplet. Since the amplitudes do not contain any information about β, γ and λ, it is necessary to use some other characteristics for the reconstruction. We propose to make use of

J. Janno, J. Engelbrecht, *Microstructured Materials: Inverse Problems*,
Springer Monographs in Mathematics,
DOI 10.1007/978-3-642-21584-1_7,

points on graphs of solitary waves that differ from the extremal ones. They can be determined in the following manner. Given a level $w_* \neq A$, one registers the time when the wave either attains this level or drops below this level. Knowing also the time when the extremum (i.e. amplitude) is reached and the velocity, it is possible to compute the relative coordinate ξ_* such that $w(\xi_*) = w_*$. This gives a definite point $P(\xi_*, w_*)$ on the graph of the wave.

In the sequel we call a problem *balanced* if the number of unknowns equals the number of equations and *unbalanced* in case the number of equations exceeds the number of unknowns. Unbalanced problems are more stable with respect to statistical errors of the data.

The amount of information available in NDE may be various depending on the possibilities to choose and measure the waves. In good cases it is possible to determine many points on the graphs and fit the parameters with these data solving an unbalanced problem. However, in any case the following principal question should be answered: what is the minimal number of points with suitable configuration that is sufficient for the unique reconstruction? This is important, because the set of all points used in inverse problem should be chosen so that it contains as a subset this minimal configuration.

Note that again it is necessary to take into consideration at least two solitary waves. This is so because under given A a single wave has insufficient number of degrees of freedom to recover all parameters. The simple consequence of this fact is that in case (1) one must specify at least two points that lie on the graphs of different waves and in case (2) one must specify at least two points on the graph of one wave and a single point on the graph of another wave. These arguments lead us to the following formulations of *balanced* inverse problems.

IP1 Let $\lambda = 0$. Given b, μ, a point $P_1(\xi_1, w_1)$ on the graph of the first wave $w[c_1]$ and a point $P_2(\xi_2, w_2)$ on the graph of the second wave $w[c_2]$, such that $w_j \neq A_j$, $j = 1, 2$, determine the pair β, γ.

IP2 Given b, μ, two different points $P_{1l}(\xi_{1l}, w_{1l})$, $l = 1, 2$, on the graph of the first wave $w[c_1]$ and a point $P_2(\xi_2, w_2)$ on the graph of the second wave $w[c_2]$, such that $w_{1l} \neq A_1$ and $w_2 \neq A_2$, determine the triplet β, γ, λ.

As we will see in the next subsection, the uniqueness of the solution of IP2 depends on the location of the points P_{1l}, $l = 1, 2$. Therefore, it makes sense additionally to formulate an *unbalanced* inverse problem that uses more information to guarantee the uniqueness independently of the location of the points. This is

IP3 Given b, μ, three different points $P_{1l}(\xi_{1l}, w_{1l})$, $l = 1, 2, 3$, on the graph of the first wave $w[c_1]$ and a point $P_2(\xi_2, w_2)$ on the graph of the second wave $w[c_2]$, such that $w_{1l} \neq A_1$ and $w_2 \neq A_2$, determine the triplet β, γ, λ.

At the end of this section we introduce some notation that will be used throughout the chapter. In our discussion both positive and negative amplitudes occur. In the former case $w \in (0, A]$ and in the latter case $w \in [A, 0)$. To unify our notation, we give to intervals of real numbers the following *generalised meaning*:

$$(d,e) = \begin{cases} \{x : d < x < e\} & \text{in case } d < e, \\ \{x : e < x < d\} & \text{in case } d > e. \end{cases}$$

Moreover, $[d,e) = (d,e) \cup \{d\}$, $(d,e] = (d,e) \cup \{e\}$ and $[d,e] = (d,e) \cup \{d;e\}$. Then it is possible to write $w \in (0,A]$ in both the case of positive and negative A.

7.1.2 Uniqueness Issues

The problem IP1 can be solved in a closed form. From (6.19) we obtain the formula for $\frac{\kappa\xi}{2}$ in terms w and A

$$\frac{\kappa\xi}{2} = \operatorname{sign}\xi \cdot \ln\left(\sqrt{\frac{A}{w}} + \sqrt{\frac{A}{w} - 1}\right).$$

This in view of the definition of κ in (6.10) yields

$$\beta c^2 - \gamma = \frac{\xi^2(c^2 - b)}{4\delta \ln^2(\sqrt{\frac{A}{w}} + \sqrt{\frac{A}{w} - 1})}.$$

Therefore, IP1 is equivalent to the following linear system for β and γ:

$$\beta c_j^2 - \gamma = \frac{\xi_j^2(c_j^2 - b)}{4\delta \ln^2(\sqrt{\frac{A_j}{w_j}} + \sqrt{\frac{A_j}{w_j} - 1})}, \qquad j = 1,2. \tag{7.2}$$

Since $c_1^2 \neq c_2^2$, this system has a regular matrix and consequently a unique solution.

Let us consider IP2 and IP3. We are going to demonstrate the method that can be used in the study of the uniqueness of the solutions of these problems. For that reason, we choose the particular case of IP2 when the points P_{1l}, $l = 1,2$, are located at different sides of the extremum point of the first wave and present the whole uniqueness proof with all details. Proofs of other uniqueness theorems concerning IP2 and IP3 are shifted to the last section of this chapter.

As in Sect. 6.2.3, let $\xi^{\pm}(w)$ stand for the inverses of the function $w(\xi)$. From the basic equation (6.11) for $w(\xi)$ and the relations (6.10) we infer the following ODE for $\xi = \xi^{\pm}$:

$$3\delta\left(\beta c^2 - \gamma\right)\xi'(w) + 2\delta^{3/2}\lambda = \left\{3\left(c^2 - b\right)w^2 - \mu w^3\right\}\left[\xi'(w)\right]^3. \tag{7.3}$$

This form of the solitary wave equation is more convenient for the study of the inverse problems.

Let us denote the triplet of unknowns of IP2 and IP3 by $S=(\beta,\gamma,\lambda)$. To emphasise the dependence of $\xi^{\pm}(w)$ on the triplet S and the velocity c, we write either $\xi^{\pm}(w)=\xi^{\pm}[S](w)$ or $\xi^{\pm}(w)=\xi^{\pm}[S,c](w)$, as necessary.

Now let us consider IP2. Assume that the points P_{1l}, $l=1,2$, are located on different sides of the extremum. Then IP2 can be rewritten in the form of the following nonlinear system for the unknown S:

$$\xi^{+}[S,c_1](w_{11})=\xi_{11}, \qquad \xi^{-}[S,c_1](w_{12})=\xi_{12}, \qquad \xi[S,c_2](w_2)=\xi_2 \tag{7.4}$$

where the function $\xi[S,c_2]$ is either $\xi^{+}[S,c_2]$ or $\xi^{-}[S,c_2]$ depending on the location of the point P_2.

We emphasise that the functions $\xi^{\pm}[S,c]$ involved in the system (7.4) are integrals of the complicated nonlinear ODE (7.3). Our aim is to remove such integrals. To this end we reduce the system (7.4) to a similar system that contains the derivatives of $\xi^{\pm}[S,c]$ instead of $\xi^{\pm}[S,c]$. It turns out that the latter system is a simple algebraic problem and does not involve any integrals. Such a reduction can be performed by means of Rolle's theorem. We will demonstrate it now.

Suppose that IP2 has two solutions $S=(\beta,\gamma,\lambda)$ and $\widetilde{S}=(\widetilde{\beta},\widetilde{\gamma},\widetilde{\lambda})$. Then, in addition to (7.4) we have the relations

$$\xi^{+}[\widetilde{S},c_1](w_{11})=\xi_{11}, \qquad \xi^{-}[\widetilde{S},c_1](w_{12})=\xi_{12}, \qquad \xi[\widetilde{S},c_2](w_2)=\xi_2. \tag{7.5}$$

From (7.4) and (7.5) we immediately obtain

$$\xi^{+}[S,c_1](w_{11})-\xi^{+}[\widetilde{S},c_1](w_{11})=0, \tag{7.6}$$

$$\xi^{-}[S,c_1](w_{12})-\xi^{-}[\widetilde{S},c_1](w_{12})=0, \tag{7.7}$$

$$\xi[S,c_2](w_2)-\xi[\widetilde{S},c_2](w_2)=0. \tag{7.8}$$

By the definition of the amplitude, $\xi^{\pm}(A)=0$. Therefore, the relations

$$\xi^{+}[S,c_1](A_1)-\xi^{+}[\widetilde{S},c_1](A_1)=0, \tag{7.9}$$

$$\xi^{-}[S,c_1](A_1)-\xi^{-}[\widetilde{S},c_1](A_1)=0, \tag{7.10}$$

$$\xi[S,c_2](A_2)-\xi[\widetilde{S},c_2](A_2)=0 \tag{7.11}$$

are also valid. Let us consider the pair of relations (7.6) and (7.9). Due to Rolle's theorem, there exist $\overline{w}_{11}\in(w_{11},A_1)$ such that

$$\xi^{+}[S,c_1]'(\overline{w}_{11})-\xi^{+}[\widetilde{S},c_1]'(\overline{w}_{11})=0. \tag{7.12}$$

Similarly, from the pairs of relations (7.7), (7.10) and (7.8), (7.11) by means of Rolle's theorem we conclude that there exist $\overline{w}_{12}\in(w_{12},A_1)$ and $\overline{w}_2\in(w_2,A_2)$ such that

$$\begin{aligned}&\xi^{-}[S,c_1]'(\overline{w}_{12})-\xi^{-}[\widetilde{S},c_1]'(\overline{w}_{12})=0,\\ &\xi[S,c_2]'(\overline{w}_2)-\xi[\widetilde{S},c_2]'(\overline{w}_2)=0.\end{aligned} \tag{7.13}$$

The equalities (7.12) and (7.13) show that the triplets S and $\widetilde{S}$ solve a system that is similar to (7.4) but contains derivatives of $\xi^{\pm}[S,c]$ instead of $\xi^{\pm}[S,c]$. Namely,

$$\left.\begin{aligned} &\xi^{+}[S,c_1]'(\overline{w}_{11}) = \xi^{+}[\widetilde{S},c_1]'(\overline{w}_{11}) = \xi'_{11}, \\ &\xi^{-}[S,c_1]'(\overline{w}_{12}) = \xi^{-}[\widetilde{S},c_1]'(\overline{w}_{12}) = \xi'_{12}, \\ &\xi[S,c_2]'(\overline{w}_2) = \xi[\widetilde{S},c_2]'(\overline{w}_2) = \xi'_2 \end{aligned}\right\} \tag{7.14}$$

where ξ'_{11}, ξ'_{12} and ξ'_2 are certain numbers.

Let us continue with the system (7.14). We point out that this is a simple algebraic problem for S and $\widetilde{S}$, because the derivatives of $\xi^{\pm}[S,c]$ solve the cubic equation (7.3). Indeed, let us consider (7.3) for $\xi'(w) = \xi^{+}[S,c_1]'(\overline{w}_{11})$ and $\xi'(w) = \xi^{+}[\widetilde{S},c_1]'(\overline{w}_{11})$. According to the first equation in (7.14), we can plug ξ'_{11} into these equations. This leads to the following pair of algebraic relations that contain S and $\widetilde{S}$:

$$3\delta(\beta c_1^2 - \gamma)\xi'_{11} + 2\delta^{3/2}\lambda = \{3(c_1^2 - b)\overline{w}_{11}^2 - \mu\overline{w}_{11}^3\}[\xi'_{11}]^3,$$
$$3\delta(\widetilde{\beta} c_1^2 - \widetilde{\gamma})\xi'_{11} + 2\delta^{3/2}\widetilde{\lambda} = \{3(c_1^2 - b)\overline{w}_{11}^2 - \mu\overline{w}_{11}^3\}[\xi'_{11}]^3.$$

Similarly, from the second and third equations in (7.14) we deduce

$$3\delta(\beta c_1^2 - \gamma)\xi'_{12} + 2\delta^{3/2}\lambda = \{3(c_1^2 - b)\overline{w}_{12}^2 - \mu\overline{w}_{11}^3\}[\xi'_{12}]^3,$$
$$3\delta(\widetilde{\beta} c_1^2 - \widetilde{\gamma})\xi'_{12} + 2\delta^{3/2}\widetilde{\lambda} = \{3(c_1^2 - b)\overline{w}_{12}^2 - \mu\overline{w}_{12}^3\}[\xi'_{12}]^3,$$
$$3\delta(\beta c_2^2 - \gamma)\xi'_{2} + 2\delta^{3/2}\lambda = \{3(c_2^2 - b)\overline{w}_{2}^2 - \mu\overline{w}_{2}^3\}[\xi'_{2}]^3,$$
$$3\delta(\widetilde{\beta} c_2^2 - \widetilde{\gamma})\xi'_{2} + 2\delta^{3/2}\widetilde{\lambda} = \{3(c_2^2 - b)\overline{w}_{2}^2 - \mu\overline{w}_{2}^3\}[\xi'_{2}]^3.$$

Pairwise subtraction of the obtained relations leads to the following 3×3 linear system for $\widetilde{S} - S$:

$$\begin{pmatrix} 3c_1^2\xi'_{11} & -3\xi'_{11} & 2\delta^{1/2} \\ 3c_1^2\xi'_{12} & -3\xi'_{12} & 2\delta^{1/2} \\ 3c_2^2\xi'_{2} & -3\xi'_{2} & 2\delta^{1/2} \end{pmatrix} \begin{pmatrix} \widetilde{\beta} - \beta \\ \widetilde{\gamma} - \gamma \\ \widetilde{\lambda} - \lambda \end{pmatrix} = \begin{pmatrix} 0 \\ 0 \\ 0 \end{pmatrix}. \tag{7.15}$$

The determinant of this system equals

$$18\,\delta^{1/2}(c_1^2 - c_2^2)\xi'_2(\xi'_{11} - \xi'_{12}). \tag{7.16}$$

The determinant is different from zero. This follows from the inequalities

$$\xi'_2 = \xi[S,c_2]'(\overline{w}_2) \neq 0, \qquad \xi'_{11} = \xi^{+}[S,c_1]'(\overline{w}_{11}) \neq 0,$$
$$\xi'_{12} = \xi^{-}[S,c_1]'(\overline{w}_{12}) \neq 0,$$

$c_1^2 \neq c_2^2$ and $\xi'_{11} \neq \xi'_{12}$. The latter inequality is valid because the quantities ξ'_{11} and ξ'_{12} have *opposite signs*. This is so because the functions $\xi^+[S, c_1]$ and $\xi^-[S, c_1]$ behind these quantities have the opposite monotonicity. (We emphasise that the inequality $\xi'_{11} \neq \xi'_{12}$ is a very delicate point of the uniqueness proof!) Consequently, the solution of the homogeneous system (7.15) is trivial. This implies that $\widetilde{S} = S$. Summing up, we have proved the following theorem.

Theorem 7.1 *Let the points P_{1l}, $l = 1, 2$, of the first wave be located on different sides of the extremum. Then the solution of* IP2 *is unique.*

In case the points P_{1l}, $l = 1, 2$, are located on a common side of the extremum, the presented proof does not work. Then the crucial inequality $\xi'_{11} \neq \xi'_{12}$ may fail because both quantities ξ'_{1l}, $l = 1, 2$, are defined as values of a common function $\xi^{\pm}[S, c_1]'$ that repeats on the interval $(0, A_1)$. This means that the uniqueness may also fail. We will give a relevant but quite technical *counter-example* for the uniqueness in Sect. 7.4.1.

Nevertheless, it is possible to prove the uniqueness in case the points $P_{1l}, l = 1, 2$, are located on a certain part of a common side of the extremum. For instance, it is possible when P_{1l}, $l = 1, 2$, stand between the inflection point and the extremum. This means that $w_{1l} \in (\frac{2A_1}{3}, A_1)$, $l = 1, 2$, because the inflection point occurs at the level $w = \frac{2A_1}{3}$ (see Theorem 6.3). In such a case the important inequality $\xi'_{11} \neq \xi'_{12}$ follows from the strict monotonicity of the derivatives of $\xi^{\pm}[S, c_1]$ between $\frac{2A_1}{3}$ and A_1. Let us formulate the corresponding uniqueness result. (A detailed proof is presented in Sect. 7.4.1.)

Theorem 7.2 *Let the points P_{1l}, $l = 1, 2$, of the first wave be located on a common side of the extremum between the inflection point and the extremum, i.e., $w_{1l} \in (\frac{2A_1}{3}, A_1)$, $l = 1, 2$. Then the solution of* IP2 *is unique.*

Finally, the solution of IP3 is unique independently of the location of the points P_{1l}, $l = 1, 2, 3$. The reason is that among these three points a pair satisfying the inequality $\xi'_{11} \neq \xi'_{12}$ always exists. This is so because the functions $\xi^{\pm}[S, c_1]'$ have only two repeating values on the interval $(0, A_1)$. The detailed proof of the following uniqueness theorem is again presented in Sect. 7.4.1.

Theorem 7.3 *The solution of* IP3 *is unique.*

7.1.3 Stability Estimates

Another important issue related to the inverse problem is the stability with respect to errors of the data. In our analysis we take two types of errors into account. The first one is related to the inaccuracy of fixing the levels of measurement of the waves w.

The second one is related to the inaccuracy of the measurement of time moments during the experiment and leads to errors in the coordinate ξ. We are going to study the impact of these errors on the solution.

We start by considering IP1. Denote the approximate levels by $\widetilde{w}_1$, $\widetilde{w}_2$ and the approximate ξ-coordinates by $\widetilde{\xi}_1, \widetilde{\xi}_2$. The components of the corresponding approximate solution is denoted by $\widetilde{\beta}$ and $\widetilde{\gamma}$. By (7.2) we immediately obtain the following relations for the difference of approximate and exact solutions:

$$\begin{aligned} \delta(\widetilde{\beta}-\beta) &= \frac{1}{4(c_2^2-c_1^2)}\big[g_2(\widetilde{w}_2,\widetilde{\xi}_2)-g_2(w_2,\xi_2) \\ &\quad - g_1(\widetilde{w}_1,\widetilde{\xi}_1)+g_1(w_1,\xi_1)\big], \\ \delta(\widetilde{\gamma}-\gamma) &= \frac{1}{4(c_2^2-c_1^2)}\big[c_2^2\big(g_2(\widetilde{w}_2,\widetilde{\xi}_2)-g_2(w_2,\xi_2)\big) \\ &\quad - c_1^2\big(g_1(\widetilde{w}_1,\widetilde{\xi}_1)-g_1(w_1,\xi_1)\big)\big] \end{aligned} \tag{7.17}$$

where

$$g_j(w,\xi) = \frac{\xi^2 l(c_j^2-b)}{\ln^2\big(\sqrt{\frac{A_j}{w}}+\sqrt{\frac{A_j}{w}-1}\big)}, \quad j=1,2. \tag{7.18}$$

Representing the involved differences of values of g in the form

$$\begin{aligned} &g_j(\widetilde{w}_j,\widetilde{\xi}_j)-g_j(w_j,\xi_j) \\ &\quad = \int_0^1 \big[g_{j,w}\big((1-t)\widetilde{w}_j+tw_j,(1-t)\widetilde{\xi}_j+t\xi_j\big)(w_j-\widetilde{w}_j) \\ &\qquad + g_{j,\xi}\big((1-t)\widetilde{w}_j+tw_j,(1-t)\widetilde{\xi}_j+t\xi_j\big)(\xi_j-\widetilde{\xi}_j)\big]\,dt \end{aligned}$$

and estimating the partial derivatives under the integral, we immediately reach the following *local Lipschitz estimate* for the solutions:

$$\begin{aligned} \max\big\{\delta|\beta-\widetilde{\beta}|;\delta|\gamma-\widetilde{\gamma}|\big\} &\le \frac{\max\{|c_1^2-b|;|c_2^2-b|\}}{4|c_1^2-c_2^2|} \\ &\quad \times \big[N_0(d,\widetilde{d})\varepsilon_\xi+\overline{N}_0(d,\widetilde{d})\varepsilon_w\big] \end{aligned} \tag{7.19}$$

where $d=(w_1,w_2,\xi_1,\xi_2)$, $\widetilde{d}=(\widetilde{w}_1,\widetilde{w}_2,\widetilde{\xi}_1,\widetilde{\xi}_2)$ are the data vectors,

$$\varepsilon_\xi = |\xi_1-\widetilde{\xi}_1|+|\xi_2-\widetilde{\xi}_2|, \qquad \varepsilon_w = |w_1-\widetilde{w}_1|+|w_2-\widetilde{w}_2| \tag{7.20}$$

are the errors of the data and

$$
\begin{aligned}
N_0(d,\widetilde{d}) &= 2\sum_{j=1}^{2} \max_{\substack{\xi\in[\xi_j,\widetilde{\xi}_j]\\ w\in[w_j,\widetilde{w}_j]}} \frac{|\xi|}{\ln^2(\sqrt{\frac{A_j}{w}}+\sqrt{\frac{A_j}{w}-1})},\\
\overline{N}_0(d,\widetilde{d}) &= \sum_{j=1}^{2} \max_{\substack{\xi\in[\xi_j,\widetilde{\xi}_j]\\ w\in[w_j,\widetilde{w}_j]}} \frac{\xi^2|A_j|}{w^2|\ln^3(\sqrt{\frac{A_j}{w}}+\sqrt{\frac{A_j}{w}-1})|\sqrt{\frac{A_j}{w}(\frac{A_j}{w}-1)}}
\end{aligned}
\tag{7.21}
$$

are the Lipschitz coefficients. Let us summarise this result in the following theorem.

Theorem 7.4 *The solutions of* IP1 *corresponding to the data vectors* d *and* $\widetilde{d}$, *respectively, satisfy the estimate* (7.19) *with the coefficients* (7.21) *and the data errors* (7.20).

The coefficients N_0 and $\overline{N}_0$ are bounded in every compact subset of the domain D_0^2 where

$$
D_0 = (0, A_1) \times (0, A_2) \times (0, \sigma_1\infty) \times (0, \sigma_2\infty)
$$

and $\sigma_j = \operatorname{sign}\xi_j$, $j = 1, 2$. Consequently, for any compact $\mathcal{D} \subset D_0^2$ and any $(d, \widetilde{d}) \in \mathcal{D}$ the estimate

$$
\max\{|\beta - \widetilde{\beta}|; |\gamma - \widetilde{\gamma}|\} \le C_{\mathcal{D}}[\varepsilon_\xi + \varepsilon_w]
$$

holds, where the coefficient $C_{\mathcal{D}}$ depends on the subset $\mathcal{D}$. In particular, if the error of the data tends to zero, i.e., $\varepsilon_\xi + \varepsilon_w \to 0$, then the error of the solution also approaches zero, i.e.,

$$
\max\{|\beta - \widetilde{\beta}|; |\gamma - \widetilde{\gamma}|\} \to 0.
$$

This shows the stability inside $\mathcal{D}$. However, we cannot extend the stability result to the whole set D_0^2 because the coefficients N_0 and $\overline{N}_0$ increase in the neighbourhood of the boundary of D_0^2.

To show what happens with the error near the boundary of D_0^2, we are going to consider two particular cases: w_1 approaches the amplitude A_1 and w_1 approaches zero.

In the first case let us choose some sequence of levels w_1^n such that $w_1^n \to A_1$ and denote the corresponding ξ-coordinates by ξ_1^n. Then $\xi_1^n \to 0$ because the measurement points approach the extremum. For the sake of simplicity assume that $w_2, \widetilde{w}_2, \xi_2, \widetilde{\xi}_2$ are fixed and the errors of the levels w_1^n are zero, i.e., $\widetilde{w}_1^n = w_1^n$ and the errors of ξ_1^n are positive and constant, i.e., $\widetilde{\xi}_1^n = \xi_1^n + \epsilon$, $\epsilon > 0$. Denote the related sequence of the approximate solutions by $(\widetilde{\beta}^n, \widetilde{\gamma}^n)$. From (7.17) we have

$$
\widetilde{\beta}^n - \beta = -\frac{1}{4\delta(c_2^2 - c_1^2)}\left[g_1(w_1^n, \xi_1^n + \epsilon) - g_1(w_1^n, \xi_1^n)\right] + K_1,
$$

$$
\widetilde{\gamma}^n - \gamma = -\frac{c_1^2}{4\delta(c_2^2 - c_1^2)}\left[g_1(w_1^n, \xi_1^n + \epsilon) - g_1(w_1^n, \xi_1^n)\right] + K_2
$$

with the constants $K_1 = \frac{1}{4\delta(c_2^2-c_1^2)}[g_2(\widetilde{w}_2,\widetilde{\xi}_2) - g_2(w_2,\xi_2)]$, $K_2 = \frac{c_2^2}{4\delta(c_2^2-c_1^2)} \times [g_2(\widetilde{w}_2,\widetilde{\xi}_2) - g_2(w_2,\xi_2)]$. In view of (7.2) and (7.18) the sequence $g_1(w_1^n,\xi_1^n)$ is also constant, namely $g_1(w_1^n,\xi_1^n) = 4\delta(\beta c_1^2 - \gamma)$. But

$$|g_1(w_1^n,\xi_1^n+\epsilon)| = \left|\frac{(\xi_1^n+\epsilon)^2(c_1^2-b)}{\ln^2(\sqrt{\frac{A_1}{w_1^n}}+\sqrt{\frac{A_1}{w_1^n}-1})}\right| \to \infty$$

because $w_1^n \to A_1$ and $\xi_1^n \to 0$. This implies that the error of the solution increases, i.e., $|\widetilde{\beta}^n - \beta| \to \infty$ and $|\widetilde{\gamma}^n - \gamma| \to \infty$. The measurements near the extrema are very sensitive with respect of the noise of the data. This result is expected because, as we saw in Sect. 7.1.1, the amplitudes do not contain any information concerning the pair β,γ.

In the second case we choose a sequence $w_1^n \to 0$. Then $|\xi_1^n| \to \infty$. As before, let $w_2,\widetilde{w}_2,\xi_2,\widetilde{\xi}_2$ be fixed. This time assume that $\widetilde{w}_1^n = w_1^n + \epsilon$ with some $\epsilon > 0$ and $\widetilde{\xi}_1^n = \xi_1^n$. Then for the approximate solution the formulas

$$\widetilde{\beta}^n - \beta = -\frac{1}{4\delta(c_2^2-c_1^2)}[g_1(w_1^n+\epsilon,\xi_1^n) - g_1(w_1^n,\xi_1^n)] + K_1,$$

$$\widetilde{\gamma}^n - \gamma = -\frac{c_1^2}{4\delta(c_2^2-c_1^2)}[g_1(w_1^n+\epsilon,\xi_1^n) - g_1(w_1^n,\xi_1^n)] + K_2$$

hold with the same constants K_1, K_2 and the constant quantities $g_1(w_1^n,\xi_1^n)$. Now we have

$$|g_1(w_1^n+\epsilon,\xi_1^n)| = \left|\frac{(\xi_1^n)^2(c_1^2-b)}{\ln^2(\sqrt{\frac{A_1}{w_1^n+\epsilon}}+\sqrt{\frac{A_1}{w_1^n+\epsilon}-1})}\right| \to \infty$$

in view of the relations $w_1^n \to 0$ and $|\xi_1^n| \to \infty$. Consequently, $|\widetilde{\beta}^n - \beta| \to \infty$ and $|\widetilde{\gamma}^n - \gamma| \to \infty$. This shows that measurements in low levels may be very sensitive with respect to the noise of the data.

A natural question arises: can we find optimal values of w_j between 0 and A_j? Let us try to find such an optimum minimising the right-hand side of the estimate (7.19) over w_1, w_2. For a sake of simplicity, assume that ε_w and ε_ξ are very small. This means that the coefficients N_0 and $\overline{N}_0$ can be computed approximately as follows (see also (7.2)):

$$N_0(d,\widetilde{d}) \approx 2\sum_{j=1}^{2}\frac{|\xi_j|}{\ln^2(\sqrt{\frac{A_j}{w_j}}+\sqrt{\frac{A_j}{w_j}-1})} = 4\sum_{j=1}^{2}\frac{\sqrt{|\frac{\delta(\beta c_j^2-\gamma)}{c_j^2-b}|}}{|\ln(\sqrt{\frac{A_j}{w_j}}+\sqrt{\frac{A_j}{w_j}-1})|},$$

$$\overline{N}_0(d,\widetilde{d}) \approx \sum_{j=1}^{2}\frac{\xi_j^2|A_j|}{w_j^2|\ln^3(\sqrt{\frac{A_j}{w_j}}+\sqrt{\frac{A_j}{w_j}-1})|\sqrt{\frac{A_j}{w_j}(\frac{A_j}{w_j}-1)}}$$

$$= 4\sum_{j=1}^{2}\frac{|\frac{\delta A_j(\beta c_j^2-\gamma)}{c_j^2-b}|}{w_j^2|\ln(\sqrt{\frac{A_j}{w_j}}+\sqrt{\frac{A_j}{w_j}-1})|\sqrt{\frac{A_j}{w_j}(\frac{A_j}{w_j}-1)}}.$$

The first coefficient N_0 increases in both $|w_1|$ and $|w_2|$ but the other coefficient $\overline{N}_0$ is a sum of products of decreasing and increasing functions of w_j. Therefore, the minimum of the main factor $N_0\varepsilon_\xi + \overline{N}_0\varepsilon_w$ in (7.19) depends on the relation of the error ε_ξ to the error ε_w. Clearly, optimal values of w_j can be found in case such a relation is a priori known.

We continue by discussing the stability of IP2. As above, we denote the vectors of exact and approximate data by d and $\widetilde{d}$, respectively. The data vectors consist of the following coordinates:

$$d = (w_{11}, w_{12}, w_2, \xi_{11}, \xi_{12}, \xi_2), \qquad \widetilde{d} = (\widetilde{w}_{11}, \widetilde{w}_{12}, \widetilde{w}_2, \widetilde{\xi}_{11}, \widetilde{\xi}_{12}, \widetilde{\xi}_2).$$

The exact and approximate solutions are the triplets β, γ, λ and $\widetilde{\beta}, \widetilde{\gamma}, \widetilde{\lambda}$, respectively. For the sake of simplicity, we consider only the case when the measurements are taken from different sides of the extremum of the first wave. Then the following theorem provides a Lipschitz estimate for the difference of the solutions.

Theorem 7.5 *Let the points P_{1l}, $l = 1, 2$, be located on different sides of the extremum. Then the solutions of* IP2 *corresponding to the data vectors d and $\widetilde{d}$, respectively, satisfy the estimate*

$$\max\{\delta|\beta - \widetilde{\beta}|; \delta|\gamma - \widetilde{\gamma}|; \delta^{3/2}|\lambda - \widetilde{\lambda}|\} \le \frac{C_*}{|c_1^2 - c_2^2|^2}\left[N(d, \widetilde{d})\varepsilon_\xi + \overline{N}(d, \widetilde{d})\varepsilon_w\right] \tag{7.22}$$

where

$$\begin{aligned} \varepsilon_\xi &= |\xi_{11} - \widetilde{\xi}_{11}| + |\xi_{12} - \widetilde{\xi}_{12}| + |\xi_2 - \widetilde{\xi}_2|, \\ \varepsilon_w &= |w_{11} - \widetilde{w}_{11}| + |w_{12} - \widetilde{w}_{12}| + |w_2 - \widetilde{w}_2|, \end{aligned} \tag{7.23}$$

are the errors of the data and C_* is a constant depending on μ, $\max\{|A_1|; |A_2|\}$, $\max\{c_1^2; c_2^2\}$. Also, the Lipschitz coefficients $N(d, \widetilde{d})$, $\overline{N}(d, \widetilde{d})$ are given by the formulas

$$\begin{aligned} N(d, \widetilde{d}) &= \frac{(1 + L(d))([Q(d, \widetilde{d})]^2 + K(\widetilde{d}))}{z_A(d, \widetilde{d})}, \\ \overline{N}(d, \widetilde{d}) &= \frac{(1 + L(d))Q(d, \widetilde{d})}{[z_A(d, \widetilde{d})]^{3/2}}\left(\frac{[Q(d, \widetilde{d})]^2}{[z_0(d, \widetilde{d})]^3} + \frac{K(\widetilde{d})}{z_0(d, \widetilde{d})}\right) \end{aligned} \tag{7.24}$$

with

$$\begin{aligned} K(d) = &\left[1 + \frac{|w_{11} - A_1|}{|\xi_{11}|} + \frac{|w_{12} - A_1|}{|\xi_{12}|} + \frac{|w_2 - A_2|}{|\xi_2|}\right] \\ &\times \left[\frac{|\xi_{11}|^3}{|w_{11} - A_1|^2} + \frac{|\xi_{12}|^3}{|w_{12} - A_1|^2} + \frac{|\xi_2|^3}{|w_2 - A_2|^2}\right], \end{aligned} \tag{7.25}$$

$$L(d) = \max\left\{\frac{|I_0[A_1](w_{1l})|}{|\xi_{1l}|}\sqrt{|A_1 - w_{1l}|}\ (l = 1, 2);\right.$$
$$\left.\frac{|I_0[A_2](w_2)|}{|\xi_2|}\sqrt{|A_2 - w_2|}\right\}, \tag{7.26}$$

$$Q(d, \widetilde{d}) = \max\left\{\frac{|\xi_{1l}|}{|I_0[A_1](w_{1l})|}\ (l = 1, 2);\ \frac{|\xi_2|}{|I_0[A_2](w_2)|};\right.$$
$$\left.\frac{|\widetilde{\xi}_{1l}|}{|I_0[A_1](\widetilde{w}_{1l})|}\ (l = 1, 2);\ \frac{|\widetilde{\xi}_2|}{|I_0[A_2](\widetilde{w}_2)|}\right\}, \tag{7.27}$$

$$z_0(d, \widetilde{d}) = \min\{|w_{11}|; |w_{12}|; |w_2|; |\widetilde{w}_{11}|; |\widetilde{w}_{12}|; |\widetilde{w}_2|\},$$
$$z_A(d, \widetilde{d}) = \min\{|w_{11} - A_1|; |w_{12} - A_1|; |w_2 - A_2|; \tag{7.28}$$
$$|\widetilde{w}_{11} - A_1|; |\widetilde{w}_{12} - A_1|; |\widetilde{w}_2 - A_2|\}$$

where the w- (and A-)dependent function $I_0(w) = I_0[A](w)$ is given by (6.37).

To prove this theorem, we have to show the stability for the equivalent nonlinear system (7.4). For this purpose, it is possible to make use of ideas already applied in the uniqueness proofs. Namely, we reduce the stability of (7.4) to the stability of a similar system containing derivatives ξ' instead of the functions ξ. In this reduction stage it is possible to apply Lagrange's mean value theorem. The stability for the obtained system containing ξ' can be proved by studying the algebraic equation (7.3) for ξ'. However, all these procedures involve rather complicated technical computations. Therefore, we shift the proof to Sect. 7.4.2.

The Lipschitz coefficients N and $\overline{N}$ of the estimate (7.22) continuously depend on $(d, \widetilde{d})$ in the domain D^2, where

$$D = (0, A_1)^2 \times (0, A_2) \times (0, \infty) \times (0, -\infty) \times (0, \sigma\infty)$$

with $\sigma = \operatorname{sign} \xi_2$. Therefore, for any compact $\mathcal{D} \subset D^2$ and any $(d, \widetilde{d}) \in \mathcal{D}$ the estimate

$$\max\{|\beta - \widetilde{\beta}|; |\gamma - \widetilde{\gamma}|; |\lambda - \widetilde{\lambda}|\} \leq C^*_{\mathcal{D}}[\varepsilon_\xi + \varepsilon_w]$$

is valid where the coefficient $C^*_{\mathcal{D}}$ depends on $\mathcal{D}$. This implies the stability, i.e.,

$$\max\{|\beta - \widetilde{\beta}|; |\gamma - \widetilde{\gamma}|; |\lambda - \widetilde{\lambda}|\} \to 0 \quad \text{as } \varepsilon_\xi + \varepsilon_w \to 0.$$

As in the case of IP1, the error of the solution becomes worse in the neighbourhood of the boundary of D^2 where the measurement levels are close to either the amplitude value or zero. However, for IP2 we do not have explicit formulas for the solution of the inverse problem and hence we cannot analyse the exact behaviour of the solutions at the boundary of D^2. Therefore, let we limit ourselves to the study of the behaviour of the right-hand side of the estimate (7.22) of the solutions near the boundary of D^2. Let us denote this right-hand side by RHS.

Firstly, we consider the behaviour of the estimate when a level approaches an amplitude. Since the amplitudes do not contain any information about the triplet S, RHS is expected to increase in such a process. Let us choose a sequence of levels w_{11}^n such that $w_{11}^n \to A_1$ and denote the corresponding ξ-coordinates by ξ_{11}^n. Then $\xi_{11}^n \to 0$. As in the case of IP1, we let the other data be fixed. This means that $w_{12}, \widetilde{w}_{12}, w_2, \widetilde{w}_2, \xi_{12}, \widetilde{\xi}_{12}, \xi_2, \widetilde{\xi}_2$ are independent of n. Further, we suppose that the errors of w_{11}^n are zero, i.e., $\widetilde{w}_{11}^n = w_{11}^n$ and the errors of ξ_{11}^n are constant and nonzero, i.e., $\widetilde{\xi}_{11}^n = \xi_{11}^n + \epsilon$, $\epsilon \neq 0$. We denote the n-dependent vectors of exact and approximate data by d^n and $\widetilde{d}^n$, respectively, and the sequence of the approximate solutions by $\widetilde{\beta}^n, \widetilde{\gamma}^n, \widetilde{\lambda}^n$. Then

$$\text{RHS} = C_5^* N\big(d^n, \widetilde{d}^n\big)|\epsilon|$$

where C_5^* is a constant independent of n and $N(d^n, \widetilde{d}^n)$ is given by the formula (7.24). We have to establish the behaviour of $N(d^n, \widetilde{d}^n)$. Due to the estimate (7.66) proved in Sect. 7.4.1, all quotients of the type $\frac{|\xi^{\pm}[S,c](w)|}{|I_0(w)|}$ in the formulas (7.26), (7.27) are bounded from below and above by positive constants independent of w. Moreover, by the relation $I_0(w) \sim -\frac{2}{\sqrt{|A|}}\sqrt{|w-A|}$ as $w \to A$ (cf. (6.37)) and (7.66), the term $|K(\widetilde{d}^n)|$ is bounded from below and above by positive constants independent of n. Now we see that only the denominator $z_A(d^n, \widetilde{d}^n)$ influences the behaviour of $N(d^n, \widetilde{d}^n)$. All other terms are bounded from below and above by positive constants independent of n. Therefore, we get the relations

$$\frac{C_6^*}{|w_{11}^n - A_1|} \le N\big(d^n, \widetilde{d}^n\big) \le \frac{C_7^*}{|w_{11}^n - A_1|}$$

where C_6^* and C_7^* are some constants. This shows that RHS increases as $n \to \infty$.

Secondly, we choose a sequence of low levels $w_{11}^n \to 0$. In this case $|\xi_{11}^n| \to \infty$. As before, $w_{12}, \widetilde{w}_{12}, w_2, \widetilde{w}_2, \xi_{12}, \widetilde{\xi}_{12}, \xi_2, \widetilde{\xi}_2$ are fixed. Suppose that the errors of the sequence of first levels are constant and differ from zero, i.e., $\widetilde{w}_{11}^n = w_{11}^n + \epsilon$ with $\epsilon \neq 0$, but the errors of the corresponding ξ-coordinates equal zero, i.e., $\widetilde{\xi}_{11}^n = \xi_{11}^n$. Under such conditions we have

$$\text{RHS} = C_8^* \overline{N}\big(d^n, \widetilde{d}^n\big)|\epsilon|$$

where the constant C_8^* is independent of n. Due to the boundedness of the quotients of the types $\frac{|\xi^{\pm}[S,c](w)|}{|I_0(w)|}$ and $|w - A|$ from below and above, all terms except for $K(\widetilde{d}^n)$ and $z_0(d^n, \widetilde{d}^n)$ in the formula of $N(d^n, \widetilde{d}^n)$ are bounded from below and above by constants independent of n. Further, by (6.14) and (7.25) we have $K(\widetilde{d}^n) \sim \text{Const}|\ln|w_{11}^n||^3$ as $n \to \infty$. Using this relation and (7.28) we deduce the estimates

$$\frac{C_9^* |\ln|w_{11}^n||^3}{|w_{11}^n|} \le \overline{N}\big(d^n, \widetilde{d}^n\big) \le \frac{C_{10}^*}{|w_{11}^n|^3}$$

for sufficiently large n with some constants C_9^* and C_{10}^*. Again, we see that RHS increases as $n \to \infty$.

Finally, we point out that the right-hand sides of both (7.19) and (7.22) increase as $c_1^2 - c_2^2 \to 0$. This supports the statement that a single solitary wave does not contain enough information to recover all coefficients.

7.2 Inverse Problems for Coupled System

7.2.1 Formulation of Inverse Problems

In this section we discuss the reconstruction of parameters of the coupled system (3.40), (3.41) by means of measurements of solitary waves. We emphasise it is realistic to measure the macro-component $w(\xi)$ of the wave, only. As we saw in Sect. 6.3, the equation for w involves the product $\vartheta = \vartheta_0 \vartheta_1$ and the quotient $\nu = \nu_1/\vartheta_0$ instead of the triplet $\vartheta_0, \vartheta_1, \nu_1$. Therefore, we concentrate on problems to determine the following six parameters: $a_0, a_1, \vartheta, \alpha, \mu$ and ν.

Inverse problems to be considered in this section have some common features with the inverse problems for the hierarchical equation treated in the previous section. For instance, a single solitary wave does not contain enough information to recover all six parameters. This is so, because the solitary wave equation (6.58) has only 4 degrees of freedom: A_0, κ, Θ and Θ_1. Another common feature is that the amplitudes do not contain enough information to reconstruct all unknowns. Indeed, from the formulas (6.47) and (6.48) in view of $w' = 0$ we see that the amplitude satisfies the equation

$$\frac{a_0 - c^2 - \frac{\vartheta}{\alpha}}{\mu} + \frac{\frac{2\vartheta}{\alpha} + 3c^2 - 3a_0}{3(c^2 - a_0)} A - \frac{\mu}{4(c^2 - a_0)} A^2 = 0. \tag{7.29}$$

From this relation we see that we may expect to recover maximally the three quantities a_0, μ and $\frac{\vartheta}{\alpha}$ from amplitudes of different waves. To determine the remaining three (in case $\nu = 0$ two) unknowns, front or rear points of the waves are to be measured, as well.

But there are essential differences between these two classes of inverse problems, too. For instance, in the present case we cannot extract some subset of parameters from measurements of amplitudes of only two waves, as in previous section. Moreover, systems occurring in the study of the uniqueness of the inverse problems for the coupled system, are of higher polynomial form. This feature complicates the analysis of these problems. Actually, the uniqueness of balanced inverse problems cannot be expected. Additionally, it should be remarked that when the nonlinearity in the micro-scale is absent, i.e., $\nu = 0$, the inverse problem cannot be solved in an explicit form. Despite having explicit formulas for $\xi^{\pm}(w)$ (6.80) and (6.81) in this case, they are too complicated to be reasonably used for the analytical solution of inverse problems.

Now we are going to formulate some inverse problems for the solitary wave equation of the macro-component of the coupled system. As in the previous section, we denote the dependence of w on the velocity c in the following manner: $w = w[c](\xi)$. Suppose that we have two solitary waves with velocities c_1 and c_2 such that $c_1^2 \neq c_2^2$. Let A_1 and A_2 stand for the amplitudes of these waves. We pose two inverse problems for this pair of waves.

IP4 Let $\nu = 0$. Given the amplitudes of both waves, k_1 different points $P_{1l}(\xi_{1l}, w_{1l})$, $l = 1, \ldots, k_1$, on the graph of the first wave $w[c_1]$ and k_2 different points $P_{2l}(\xi_{2l}, w_{2l})$, $l = 1, \ldots, k_2$, on the graph of the second wave $w[c_2]$, such that $w_{1l} \neq A_1$ and $w_{2l} \neq A_2$, determine a_0, a_1, ϑ, α and μ.

IP5 Given the amplitudes of both waves, k_1 different points $P_{1l}(\xi_{1l}, w_{1l})$, $l = 1, \ldots, k_1$, on the graph of the first wave $w[c_1]$ and k_2 different points $P_{2l}(\xi_{2l}, w_{2l})$, $l = 1, \ldots, k_2$, on the graph of the second wave $w[c_2]$, such that $w_{1l} \neq A_1$ and $w_{2l} \neq A_2$, determine a_0, a_1, ϑ, α, μ and ν.

As we will see in the next subsection, the number of points P_{il} in these problems must be much higher than the dimension of the vector of unknowns in order to guarantee the uniqueness. Another approach relies on the usage of more waves and the corresponding amplitudes in order to avoid the measurement of large numbers of front or rear points of the waves. Namely, suppose that we can measure k waves with velocities $c_1, \ldots, c_k$ such that $c_i^2 \neq c_j^2$ for $i = j$. Let $A_1, \ldots, A_k$ be the amplitudes of these waves. Assume that $A_1, \ldots, A_k$ are also different. We pose the following problems.

IP6 Let $\nu = 0$ and $k \geq 2$. Given the amplitudes $A_1, \ldots, A_k$ and two points $P_j(\xi_j, w_j)$, $j = 1, 2$, located on the waves $w[c_1]$ and $w[c_2]$, respectively, such that $w_1 \neq A_1$ and $w_2 \neq A_2$, determine a_0, a_1, ϑ, α and μ.

IP7 Let $k \geq 2$. Given the amplitudes $A_1, \ldots, A_k$, two different points $P_{1l}(\xi_{1l}, w_{1l})$, $l = 1, 2$, on the graph of the first wave $w[c_1]$ and a point $P_2(\xi_2, w_2)$ on the graph of the second wave $w[c_2]$ such that $w_{11} \neq A_1$, $w_{12} \neq A_1$ and $w_2 \neq A_2$, determine a_0, a_1, ϑ, α, μ and ν.

The inverse problems are much simpler and require remarkably less information if some of the physical parameters are a priori unknown. For instance, we consider the case when the quantities a_0 and μ be given in advance. (Recall that these parameters are related to the potential energy terms U_X^2, U_X^3 and the density ρ_0 (cf. (3.16), (3.17), (3.32) and (3.39)).) Then we are able to prove uniqueness for the following balanced problems.

IP8 Let $\nu = 0$. Given a_0, μ, the amplitude A_1 and two points $P_j(\xi_j, w_j)$, $j = 1, 2$, located on the waves $w[c_1]$ and $w[c_2]$, respectively, such that $w_1 \neq A_1$ and $w_2 \neq A_2$, determine a_1, ϑ and α.

IP9 Given a_0, μ, the amplitude A_1, two different points $P_{1l}(\xi_{1l}, w_{1l}), l = 1, 2$, on the graph of the first wave $w[c_1]$ and a point $P_2(\xi_2, w_2)$ on the graph of the second wave $w[c_2]$ such that $w_{11} \neq A_1$, $w_{12} \neq A_1$ and $w_2 \neq A_2$, determine a_1, ϑ, α and ν.

Finally, we remark that a supplement of the posed problems with a hypothetical measurement of the micro-component χ enables ϑ_0 to be reconstructed as well.

Indeed, having the amplitude $A_{1,\mathrm{mirco}}$ of $\chi[c_1]$, from (6.43) we obtain

$$\vartheta_0 = \frac{(c_1^2 - a_0)A_1 - \frac{\mu}{2}A_1^2}{A_{1,\mathrm{micro}}} \tag{7.30}$$

and in the presence of $a_0, a_1, \vartheta, \alpha, \mu$ and ν we can determine the whole vector of original coefficients of the coupled system (3.40), (3.41), including ϑ_1 and ν_1.

7.2.2 Uniqueness Issues

The structure of this subsection recalls that of Sect. 7.1.2. We will present a detailed and commented uniqueness proof only for IP4. The purpose is to demonstrate the method. Uniqueness theorems for other inverse problems are formulated, too, but their proofs in more compact form are shifted to Sect. 7.4.3.

In the study of the inverse problems for the coupled system it is convenient to use an equation for the inverses $\xi = \xi^{\pm}$ of the solitary function $w(\xi)$. In view of (6.46) this equation reads

$$\begin{aligned} &3\delta(c^2 - a_1)\left[c^2 - a_0 - \mu w\right]^2 \xi'(w) - 2\delta^{3/2}\nu\left[c^2 - a_0 - \mu w\right]^3 \\ &= \left\{-3\alpha\left[(c^2 - a_0)w - \frac{\mu}{2}w^2\right]^2 - 3\vartheta(c^2 - a_0)w^2 + 2\vartheta\mu w^3\right\}\left[\xi'(w)\right]^3. \end{aligned} \tag{7.31}$$

In the proofs we will use the *generic notation* S for vectors of unknowns. As in the case of the hierarchical equation, the dependence of $w(\xi)$ and $\xi^{\pm}$ on S and the velocity c is indicated inside the square brackets, i.e., $w = w[S, c](\xi)$ and $\xi^{\pm} = \xi^{\pm}[S, c](w)$.

In proofs of uniqueness results for the inverse problems for the coupled system it is possible to combine the method of mean value theorems from the previous section with the method of vanishing polynomial coefficients used in the study of inverse problems in the linear models. We will show it below.

Let us consider IP4. We are going to prove that the solution of this problem is unique in case $k_1 = k_2 = 4$. But first of all, let us make some additional assumptions concerning the data of this problem that do not restrict the generality. They are related to the symmetry of the solitary waves with respect to the amplitude point in case $\nu = 0$. Namely, it is natural to assume that all the points of a particular wave P_{il} with $i \in \{1; 2\}$ have different levels w_{il}, i.e.,

$$w_{1l_1} \neq w_{1l_2} \quad \text{and} \quad w_{2l_1} \neq w_{2l_2} \quad \text{for any } l_1 \neq l_2. \tag{7.32}$$

Indeed, otherwise the set of data contains redundant measurements: some points are pairwise symmetric. Moreover, we can reflect all the measured points to the rear side of the wave. This means that IP4 with $k_1 = k_2 = 4$ is equivalent to the following system for $S = (a_0, a_1, \vartheta, \alpha, \mu)$:

$$\xi^{+}[S, c_i](w_{il}) = \xi_{il}, \quad l = 1, \ldots, 4, \ i = 1, 2, \tag{7.33}$$

$$\xi^{+}[S, c_i](A_i) = 0, \quad i = 1, 2. \tag{7.34}$$

Let the levels be ordered so that their distance form the amplitude increases, i.e., $w_{i1} \in (w_{i2}, A_i)$, $w_{i2} \in (w_{i3}, A_i)$, $w_{i3} \in (w_{i4}, A_i)$ where $i = 1, 2$. Suppose that IP4 has another solution $\widetilde{S} = (\widetilde{a}_0, \widetilde{a}_1, \widetilde{\vartheta}, \widetilde{\alpha}, \widetilde{\mu})$, too. Then

$$\xi^+[\widetilde{S}, c_i](w_{il}) = \xi_{il}, \quad l = 1, \ldots, 4, \ i = 1, 2, \tag{7.35}$$

$$\xi^+[\widetilde{S}, c_i](A_i) = 0, \quad i = 1, 2. \tag{7.36}$$

Using Rolle's theorem we see that there exist $\overline{w}_{i1} \in (w_{i1}, A_i)$, $\overline{w}_{i2} \in (w_{i2}, w_{i1})$, $\overline{w}_{i3} \in (w_{i3}, w_{i2})$, $\overline{w}_{i4} \in (w_{i4}, w_{i3})$, $i = 1, 2$, such that

$$\xi^+[S, c_i]'(\overline{w}_{il}) = \xi^+[\widetilde{S}, c_i]'(\overline{w}_{il}) =: \xi'_{il}, \quad l = 1, \ldots, 4, \ i = 1, 2. \tag{7.37}$$

Plugging the pairs $(w, \xi'(w)) = (\overline{w}_{il}, \xi^+[S, c_i]'(\overline{w}_{il}))$, $i = 1, 2$, and $(w, \xi'(w)) = (\overline{w}_{il}, \xi^+[\widetilde{S}, c_i]'(\overline{w}_{il}))$, $i = 1, 2$, into (7.31) and setting $\nu = 0$, the system (7.37) takes the following form of the pair of two algebraic systems:

$$\begin{aligned} &3\delta\big(c_i^2 - a_1\big)\big[c_i^2 - a_0 - \mu\overline{w}_{il}\big]^2 \\ &\quad = \left\{-3\alpha\left[\big(c_i^2 - a_0\big)\overline{w}_{il} - \frac{\mu}{2}\overline{w}_{il}^2\right]^2 - 3\vartheta\big(c_i^2 - a_0\big)\overline{w}_{il}^2 + 2\vartheta\mu\overline{w}_{il}^3\right\}\big[\xi'_{il}\big]^2, \\ &l = 1, \ldots, 4, \ i = 1, 2, \end{aligned} \tag{7.38}$$

$$\begin{aligned} &3\delta\big(c_i^2 - \widetilde{a}_1\big)\big[c_i^2 - \widetilde{a}_0 - \widetilde{\mu}\overline{w}_{il}\big]^2 \\ &\quad = \left\{-3\widetilde{\alpha}\left[\big(c_i^2 - \widetilde{a}_0\big)\overline{w}_{il} - \frac{\widetilde{\mu}}{2}\overline{w}_{il}^2\right]^2 - 3\widetilde{\vartheta}\big(c_i^2 - \widetilde{a}_0\big)\overline{w}_{il}^2 + 2\widetilde{\vartheta}\widetilde{\mu}\overline{w}_{il}^3\right\}\big[\xi'_{il}\big]^2, \\ &l = 1, \ldots, 4, \ i = 1, 2. \end{aligned} \tag{7.39}$$

Note that the system (7.38) for S is of polynomial type with the degree 3: the terms of the highest order are $a_1 a_0^2$, $a_1 a_0 \mu$ and $a_1 \mu^2$. This is the reason why the uniqueness in the balanced case cannot be expected. It is necessary to increase the number of equations and related measurements. We will show that the chosen number of equations $8 = 4 + 4$ of the form (7.38) plus 2 amplitude relations is sufficient to achieve the uniqueness.

Let us deal with the systems (7.38) and (7.39). We eliminate the quantities ξ'_{il} from these systems. To this end, we multiply (7.38) by

$$-3\widetilde{\alpha}\left[\big(c_i^2 - \widetilde{a}_0\big)\overline{w}_{il} - \frac{\widetilde{\mu}}{2}\overline{w}_{il}^2\right]^2 - 3\widetilde{\vartheta}\big(c_i^2 - \widetilde{a}_0\big)\overline{w}_{il}^2 + 2\widetilde{\vartheta}\widetilde{\mu}\overline{w}_{il}^3,$$

(7.39) by

$$-3\alpha\left[\big(c_i^2 - a_0\big)\overline{w}_{il} - \frac{\mu}{2}\overline{w}_{il}^2\right]^2 - 3\vartheta\big(c_i^2 - a_0\big)\overline{w}_{il}^2 + 2\vartheta\mu\overline{w}_{il}^3$$

and subtract. After dividing by $\overline{w}_{il}^2 \neq 0$ we obtain the relations

$$\begin{aligned}
&(c_i^2 - a_1)\big[c_i^2 - a_0 - \mu \overline{w}_{il}\big]^2 \\
&\quad \times \left\{ 3\widetilde{\alpha}\left[c_i^2 - \widetilde{a}_0 - \frac{\widetilde{\mu}}{2}\overline{w}_{il}\right]^2 + 3\widetilde{\vartheta}\left(c_i^2 - \widetilde{a}_0\right) - 2\widetilde{\vartheta}\,\widetilde{\mu}\overline{w}_{il} \right\} \\
&\quad - \left(c_i^2 - \widetilde{a}_1\right)\big[c_i^2 - \widetilde{a}_0 - \widetilde{\mu}\overline{w}_{il}\big]^2 \\
&\quad \times \left\{ 3\alpha\left[c_i^2 - a_0 - \frac{\mu}{2}\overline{w}_{il}\right]^2 + 3\vartheta\left(c_i^2 - a_0\right) - 2\vartheta\,\mu\overline{w}_{il} \right\} = 0, \\
&l = 1, \ldots, 4, \ i = 1, 2.
\end{aligned} \tag{7.40}$$

These relations show that $\overline{w}_{il}, l = 1, \ldots, 4, i = 1, 2$, are roots of the following polynomials of the 4th degree:

$$\begin{aligned}
\mathcal{P}_{4,i}(w) = {} & (c_i^2 - a_1)\big[c_i^2 - a_0 - \mu w\big]^2 \\
& \times \left\{ 3\widetilde{\alpha}\left[c_i^2 - \widetilde{a}_0 - \frac{\widetilde{\mu}}{2}w\right]^2 + 3\widetilde{\vartheta}\left(c_i^2 - \widetilde{a}_0\right) - 2\widetilde{\vartheta}\,\widetilde{\mu}w \right\} \\
& - \left(c_i^2 - \widetilde{a}_1\right)\big[c_i^2 - \widetilde{a}_0 - \widetilde{\mu}w\big]^2 \\
& \times \left\{ 3\alpha\left[c_i^2 - a_0 - \frac{\mu}{2}w\right]^2 + 3\vartheta\left(c_i^2 - a_0\right) - 2\vartheta\,\mu w \right\}, \quad i = 1, 2.
\end{aligned}$$

Further, let us take into consideration the amplitudes, too. Equations (7.34) and (7.36) are equivalent to

$$w[S, c_i](0) = w[\widetilde{S}, c_i](0) = A_i, \quad i = 1, 2.$$

Moreover, $w' = 0$ is zero at the amplitude points. Using these relations in (6.46) we obtain

$$\begin{aligned}
&3\alpha\left[c_i^2 - a_0 - \frac{\mu}{2}A_i\right]^2 + 3\vartheta\left(c_i^2 - a_0\right) - 2\vartheta\,\mu A_i = 0, \quad i = 1, 2, \\
&3\widetilde{\alpha}\left[c_i^2 - \widetilde{a}_0 - \frac{\widetilde{\mu}}{2}A_i\right]^2 + 3\widetilde{\vartheta}\left(c_i^2 - \widetilde{a}_0\right) - 2\widetilde{\vartheta}\,\widetilde{\mu}A_i = 0, \quad i = 1, 2.
\end{aligned} \tag{7.41}$$

This shows that $A_i, i = 1, 2$, are the roots of $\mathcal{P}_{4,i}, i = 1, 2$, too. Now we see that both polynomials of fourth degree $\mathcal{P}_{4,i},\ i = 1, 2$, have 5 different roots. (For $i \in \{1; 2\}$ the numbers $\overline{w}_{il},\ l = 1, \ldots, 4$, and A_i are all different from each other.) Such a situation is possible only when the polynomials $\mathcal{P}_{4,i},\ i = 1, 2$, are trivial, i.e., have zero coefficients.

In order to complete the proof of the uniqueness, we have to set the coefficients of $\mathcal{P}_{4,i}$ to zero and deduce the desired equality $\widetilde{S} = S$ from the obtained equations.

In particular, the coefficients of the term w^4 in $\mathcal{P}_{4,i}$, $i=1,2$, yield the equations

$$3(c_i^2 - a_1)\mu^2\widetilde{\alpha}\frac{\widetilde{\mu}^2}{4} - 3(c_i^2 - \widetilde{a}_1)\widetilde{\mu}^2\alpha\frac{\mu^2}{4} = 0, \quad i = 1, 2.$$

In view of the inequalities $\mu, \widetilde{\mu} \neq 0$ (they are necessary for the existence of the solitary waves, by Lemma 6.3), from these equations we deduce the following linear homogeneous 2×2 system for the quantities $\widetilde{\alpha} - \alpha$ and $\widetilde{a}_1\alpha - a_1\widetilde{\alpha}$:

$$c_i^2(\widetilde{\alpha} - \alpha) + \widetilde{a}_1\alpha - a_1\widetilde{\alpha} = 0, \quad i = 1, 2.$$

Since $c_1^2 \neq c_2^2$, this system is regular. Thus, the solution is $\widetilde{\alpha} - \alpha = \widetilde{a}_1\alpha - a_1\widetilde{\alpha} = 0$. This in view of $\alpha \neq 0$ (see (3.43)) implies that

$$\widetilde{\alpha} = \alpha, \qquad \widetilde{a}_1 = a_1. \tag{7.42}$$

Further, the zero-order terms of in $\mathcal{P}_{4,i}$, $i=1,2$, provide the equations

$$\begin{aligned} &3(c_i^2 - a_1)(c_i^2 - a_0)^2(c_i^2 - \widetilde{a}_0)\{\widetilde{\alpha}(c_i^2 - \widetilde{a}_0) + \widetilde{\vartheta}\} \\ &\quad - 3(c_i^2 - \widetilde{a}_1)(c_i^2 - \widetilde{a}_0)^2(c_i^2 - a_0)\{\alpha(c_i^2 - a_0) + \vartheta\} = 0, \quad i = 1, 2. \end{aligned}$$

They are simplified substituting $\widetilde{\alpha}$ by α and $\widetilde{a}_1$ by a_1 and dividing by the factor $3(c_i^2 - a_1)(c_i^2 - a_0)(c_i^2 - \widetilde{a}_0)$ (this is different from zero due to Lemma 6.3). The result is the linear system

$$c_i^2(\widetilde{\vartheta} - \vartheta) + \widetilde{a}_0\vartheta - a_0\widetilde{\vartheta} = 0, \quad i = 1, 2.$$

The solution of this system is $\widetilde{\vartheta} - \vartheta = \widetilde{a}_0\vartheta - a_0\widetilde{\vartheta} = 0$. Due to $\vartheta \neq 0$ (cf. (3.43)), this yields

$$\widetilde{\vartheta} = \vartheta, \qquad \widetilde{a}_0 = a_0. \tag{7.43}$$

Finally, by virtue of (7.42) and (7.43), the coefficients of the first-order terms w in $\mathcal{P}_{4,i}$, $i=1,2$, have the form

$$(c_i^2 - a_1)(c_i^2 - a_0)^2[3\alpha(c_i^2 - a_0) + 4\vartheta](\widetilde{\mu} - \mu) = 0, \quad i = 1, 2.$$

Dividing by $3(c_i^2 - a_1)(c_i^2 - a_0)^2$ and subtracting we obtain $\alpha(c_1^2 - c_2^2)(\widetilde{\mu} - \mu) = 0$. Since $\alpha \neq 0$, this implies that $\widetilde{\mu} = \mu$. Consequently, $\widetilde{S} = S$ and we have proved the following theorem.

Theorem 7.6 *Let* $k_1 = k_2 = 4$ *and let* (7.32) *hold. Then the solution of* IP4 *is unique.*

Uniqueness of solutions of IP5–IP9 are proved in a similar manner. The proofs are even more technical and can be found in Sect. 7.4.3. We give only formulations of these results here.

Theorem 7.7 *Let* $k_1 = k_2 = 16$. *Then the solution of* IP5 *is unique.*

Theorem 7.8 *Let* $k = 5$. *Then the solution of* IP6 *is unique.*

Theorem 7.9 *Let* $k = 5$ *and let the points* $P_{1l}, l = 1, 2$, *be located on different sides of the extremum. Then the solution of* IP7 *is unique.*

Theorem 7.10 *The solution of* IP8 *is unique. In particular, the given amplitude* A_1 *determines the ratio* $\frac{\vartheta}{\alpha}$ *by the simple explicit formula*

$$\frac{\vartheta}{\alpha} = -\frac{(c_1^2 - a_0 - \frac{\mu}{2}A_1)^2}{c_1^2 - a_0 - \frac{2\mu}{3}A_1}. \tag{7.44}$$

Theorem 7.11 *Let the points* P_{1l}, $l = 1, 2$, *be located on different sides of the extremum. Then the solution of* IP9 *is unique and* A_1 *determines the ratio* $\frac{\vartheta}{\alpha}$ *by the formula* (7.44).

7.3 Methods of Solution of Inverse Problems

7.3.1 Minimisation of Cost Functional

The most general method of solving inverse problems for solitary waves is based on least squares fitting. Let us consider an inverse problem whose (in general overdetermined) data consist of measurements of n waves $w[c_i]$, $i = 1, \ldots, n$, with velocities c_i. More precisely, let the data contain the amplitudes $A_1, \ldots, A_n$ of these waves and a certain set of points $P_{ij}(\xi_{ij}, w_{ij})$, $j = 1, \ldots, i_j$, of the waves $w[c_i]$ where $i = 1, \ldots, k$ and $k \leq n$. As before, let S stand for the vector of parameters to be determined in the inverse problem. We denote the admissible set of the solutions of the inverse problem by $\mathcal{S}$.

Further, as before, let $w[S, c](\xi)$ stand for the S- and c-dependent solitary wave function. We emphasise that $w[S, c]$ is the solution of the forward problem: given S, it solves the differential equation (6.9) or (6.46), depending on the model under consideration. The amplitude of the wave function equals $w[S, c](0)$.

We try to fit the data with the solution of the forward problem. To this end, we construct the least squares *cost functional*

$$J(S) = \sum_{i=1}^{n} \{w[S, c_i](0) - A_i\}^2 + \sum_{i=1}^{k} \sum_{j=1}^{i_j} \{w[S, c_i](\xi_{ij}) - w_{ij}\}^2.$$

The *quasi-solution* of the inverse problem is the vector of parameters $S^\dagger$ such that

$$S^\dagger = \arg\min_{S \in \mathcal{S}} J(S).$$

The functional $J(S)$ can be minimised by iteration techniques involving sequential solutions of ODEs (6.9) or (6.46). Various well-known minimisation techniques can be used, for instance genetic algorithms [51] or gradient-type methods [2, 47]. A good overview of contemporary methods can be found in [46].

7.3.2 Application of Series Expansion. Linearisation

For IP2 simpler methods can be constructed that avoid the solution of ODE-s. This is possible because the series representation (6.38) is available for the inverses of the solution of the related ODE (6.9).

From (6.38) we obtain

$$\xi(w) = \frac{1}{\kappa} f[\Theta](w) \quad \text{with } f[\Theta](w) = d_0 I_0(w) + \sum_{i=1}^{\infty} d_i \Theta^i I_i(w) \tag{7.45}$$

where $\kappa = \sqrt{\frac{c^2-b}{\delta(\beta c^2-\gamma)}}$, $\Theta = -2[\frac{c^2-b}{\beta c^2-\gamma}]^{3/2}\frac{\lambda}{\mu}$ and the sequence $I_i(w)$ is given by (6.37). The function $f[\Theta]$ has two branches: $f^+[\Theta]$ and $f^-[\Theta]$ generated by the sequences of d that start from the initial values $d_0 = -1$ and $d_0 = 1$, respectively (cf. (6.35)). The branches ξ^+ and ξ^- of ξ correspond to the branches $f^+[\Theta]$ and $f^-[\Theta]$ of $f[\Theta]$, respectively.

According to the definitions of κ and Θ, we introduce the quantities

$$\kappa_j = \sqrt{\frac{c_j^2-b}{\delta(\beta c_j^2-\gamma)}}, \qquad \Theta_j = -2\left[\frac{c_j^2-b}{\beta c_j^2-\gamma}\right]^{3/2}\frac{\lambda}{\mu}, \qquad j=1,2. \tag{7.46}$$

They yield further useful relations:

$$\lambda = -\frac{\mu\Theta_1}{2\delta^{3/2}\kappa_1^3} = -\frac{\mu\Theta_2}{2\delta^{3/2}\kappa_2^3}, \qquad \kappa_j = -\frac{1}{\sqrt{\delta}}\left[\frac{\mu\Theta_j}{2\lambda}\right]^{1/3}, \quad j=1,2. \tag{7.47}$$

Assume that the points P_{1l}, $l = 1, 2$, of the first wave are located on different sides of the extremum. Then, due to (7.45) and (7.46), we can write IP2 in the form of the following system:

$$f^+[\Theta_1](w_{11}) - \xi_{11}\kappa_1 = 0, \qquad f^-[\Theta_1](w_{12}) - \xi_{12}\kappa_1 = 0, \tag{7.48}$$

$$f^\sigma[\Theta_2](w_2) - \xi_2\kappa_2 = 0 \tag{7.49}$$

where $\sigma \in \{+; -\}$ depends on the side of the measurement of P_2. The first two equations (7.48) make an independent subsystem for κ_1 and Θ_1. Therefore, we can compose the following algorithm for the reconstruction of $S = (\beta, \gamma, \lambda)$:

(1) solve the 2×2 nonlinear subsystem (7.48) for κ_1 and Θ_1;
(2) using κ_1 and Θ_1 compute the parameter

$$\lambda = -\frac{\mu \Theta_1}{2\delta^{3/2}\kappa_1^3}$$

and express κ_2 in terms of Θ_2 by means of the formula

$$\kappa_2 = -\frac{1}{\sqrt{\delta}}\left[\frac{\mu\Theta_2}{2\lambda}\right]^{1/3};$$

(3) substitute the obtained formula for κ_2 into (7.49) and solve the resulting equation

$$f^{\sigma}[\Theta_2](w_2) + \xi_2 \frac{1}{\sqrt{\delta}}\left[\frac{\mu}{2\lambda}\right]^{1/3} \Theta_2^{1/3} = 0 \tag{7.50}$$

for Θ_2;
(4) compute β and γ solve the linear system

$$\beta_1 c_1^2 - \gamma_1 = \left(c_j^2 - b\right)\left[\frac{2\lambda}{\mu\Theta_j}\right]^{3/2}, \quad j = 1, 2, \tag{7.51}$$

deduced from the right-hand relations in (7.46).

The most difficult steps of this algorithm are the solution of the 2×2 system of (7.48) and the single equation (7.50). Note that the algorithm does not contain integration of ODE-s any more. The nonlinear system (7.48) and equation (7.50) can be effectively solved by means of Newton-type methods. One can even use the method of secants for the system (7.48) because it contains sums of functions of single variables Θ_1 and κ_1.

In practical computations of values of $f[\Theta]$, the series (7.45) can be truncated. The simplest procedures occur in the case of *linear approximation*

$$f[\Theta](w) \approx d_0 I_0(w) + d_1 \Theta I_1(w).$$

Then $f^{\pm}[\Theta] \approx \mp I_0(w) + \frac{1}{2}\Theta I_1(w)$ and the system (7.48) becomes linear:

$$\begin{aligned} \frac{1}{2} I_1(w_{11})\Theta_1 + \xi_{11}\kappa_1 &\approx -I_0(w_{11}), \\ \frac{1}{2} I_1(w_{12})\Theta_1 + \xi_{12}\kappa_1 &\approx I_0(w_{12}). \end{aligned} \tag{7.52}$$

The determinant of this system is different from zero because $I_1(w_{1l}) = \frac{w_{1l}}{A_1} - 1 < 0$, $l = 1, 2$, and $\xi_{11} > 0$, $\xi_{12} < 0$, where ξ_{1l}, $l = 1, 2$, are the values of the functions ξ^+ and ξ^-, respectively. In order to linearise (7.50), we cube it: $\{f^{\sigma}[\Theta_2](w_2)\}^3 = -\frac{\xi_2^3 \mu}{2\delta^{3/2}\lambda}\Theta_2$ and make the linear approximation for the cubed series in the left-hand side:

$$\left\{f^{\sigma}[\Theta_2](w_2)\right\}^3 \approx d_0\left[I_0(w_2)\right]^3 + 3d_1\left[I_0(w_2)\right]^2 I_1(w_2)\Theta_2.$$

This yields the explicit solution of (7.50):

$$\Theta_2 \approx -\frac{d_0[I_0(w_2)]^3}{3d_1[I_0(w_2)]^2 I_1(w_2) + \frac{\xi_2^3 \mu}{2\delta^{3/2}\lambda}}.$$

Here $d_1 = -\frac{1}{2}$ and $d_0 = 1$ when $\sigma = -$ and $d_0 = -1$ when $\sigma = +$.

7.3.3 Numerical Examples

We tested the sensitivity of the solutions of the problems IP1, IP2, IP4–IP9 with respect to the noise of the data. In this subsection we present and analyse the obtained results. Since IP3 is a simple extension of IP2, we will not give separate results for this problem. Actually, results for IP3 are very much the same as for IP2.

In all numerical examples we took the same basic parameters as in the linear case, i.e., $a_0 = 100$, $a_1 = 1$, $\alpha = 10^{-4}$, $\vartheta = 0.002$ (coupled system) and $b = 80$, $\beta = \gamma = 2 \times 10^5$ (hierarchical equation). In addition, we chose the following nonlinearity parameters in the coupled system: $\mu = 1$, $\nu = 100$. Then the corresponding microscale nonlinearity parameter of the hierarchical equation is $\lambda = 4 \times 10^6$ (cf. (3.45)). The geometrical parameter δ was taken equal to 10^{-4}.

In all two-wave problems (IP1, IP2, IP4, IP5, IP8, IP9) we used the waves with the velocities $c_1 = \sqrt{85}$ and $c_2 = \sqrt{98}$. According to Theorems 7.8 and 7.9 we set $k = 5$ in IP6 and IP7 and used the velocities $c_j = \sqrt{85 + 13(j-1)/4}$, $j = 1, \dots, 5$. Such choices of parameters and velocities are in an accordance with the existence conditions of the solitary waves deduced in Chap. 6.

The synthetic data for the inverse problems were constructed in the following manner. The amplitudes of waves were computed by the explicit formula (6.10) or (6.73). The levels of points of measurements were taken equal to the half amplitudes in all two- and three-point problems, namely $w_1 = \frac{A_1}{2}$, $w_2 = \frac{A_2}{2}$ in IP1, IP6, IP8, and $w_{1l} = \frac{A_1}{2}$, $l = 1, 2$, $w_2 = \frac{A_2}{2}$ in IP2, IP7, IP9. In the latter problems P_{11} and P_{12} were taken from different sides of extrema. In IP4 we set $k_1 = k_2 = 4$, according to Theorem 7.6, and chose the levels $w_{jl} = (\frac{1}{4} + \frac{l-1}{6})A_j$, $l = 1, \dots, 4$, $j = 1, 2$. Similarly, in IP5 we put $k_1 = k_2 = 16$, due to Theorem 7.7, and took the levels $w_{j,2s-1} = w_{j,2s} = (\frac{1}{6} + \frac{2(s-1)}{21})A_j$, $s = 1, \dots, 8$, $j = 1, 2$. Moreover, in IP5 the points P_{jl} with odd and even l were taken from front and rear sides of the waves, respectively. The related exact ξ-coordinates of the points P_j and P_{jl} we obtained by solving the ordinary differential equations governing the solitary wave processes. Finally, the amplitudes and ξ-coordinates were perturbed:

$$A_j^\epsilon = A_j(1 + R_{A_j}\epsilon), \qquad \xi_j^\epsilon = \xi_j(1 + R_{\xi_j}\epsilon), \qquad \xi_{jl}^\epsilon = \xi_{jl}(1 + R_{\xi_{jl}}\epsilon)$$

where R_{A_j}, R_{ξ_j} and $R_{\xi_{jl}}$ are random numbers on the interval $[-1, 1]$ and ϵ is a given relative noise level. Summing up, the noisy amplitudes A_j^ϵ and the noisy points

Table 7.1 Relative errors of solution of (7.1)

ϵ	$\lvert\frac{b^\epsilon - b}{b}\rvert$	$\lvert\frac{\mu^\epsilon - \mu}{\mu}\rvert$
0.01%	0.0036%	0.0015%
0.1%	0.03%	0.02%
1%	0.2%	0.3%

Table 7.2 Relative errors in IP1

ϵ	$\lvert\frac{\beta^\epsilon - \beta}{\beta}\rvert$	$\lvert\frac{\gamma^\epsilon - \gamma}{\gamma}\rvert$
0.01%	0.005%	0.14%
0.1%	0.08%	1.8%
1%	0.53%	26%

Table 7.3 Relative errors in IP2

ϵ	$\lvert\frac{\beta^\epsilon - \beta}{\beta}\rvert$	$\lvert\frac{\gamma^\epsilon - \gamma}{\gamma}\rvert$	$\lvert\frac{\lambda^\epsilon - \lambda}{\lambda}\rvert$
0.01%	0.008%	0.44%	0.015%
0.1%	0.11%	3.8%	0.34%
1%	0.78%	31%	4.4%

Table 7.4 Relative errors in IP4

ϵ	$\lvert\frac{a_0^\epsilon - a_0}{a_0}\rvert$	$\lvert\frac{a_1^\epsilon - a_1}{a_1}\rvert$	$\lvert\frac{\alpha^\epsilon - \alpha}{\alpha}\rvert$	$\lvert\frac{\vartheta^\epsilon - \vartheta}{\vartheta}\rvert$	$\lvert\frac{\mu^\epsilon - \mu}{\mu}\rvert$
0.01%	0.05%	0.11%	0.004%	0.35%	0.03%
0.1%	0.68%	1.5%	0.04%	6.1%	0.31%
1%	8.4%	16%	0.20%	88%	5.0%

$P_j^\epsilon(\xi_j^\epsilon, w_j)$, $P_{jl}^\epsilon(\xi_{jl}^\epsilon, w_{jl})$ formed the synthetic data for the inverse problems under consideration.

The pair b, μ in the hierarchical equation is determined from the linear system (7.1). This system is very good from the point of view of the accuracy. The results are presented in Table 7.1.

The problem IP1 also consists of a simple linear system (7.2) for the pair β, γ. The numerical results presented in Table 7.2 show that this system is much worse from the point of view of the accuracy than (7.1).

To solve IP2, we made use of the method of series expansion described in Sect. 7.3.2. The results are presented in Table 7.3.

Numerical computations concerning inverse problems for solitary waves in coupled system have been performed mainly by Sertakov [63]. We re-scaled the data and results of Sertakov's thesis and computed some additional results for IP8 and IP9. The problems IP4–IP9 were solved by the minimisation of cost functionals (Sect. 7.3.1). The minimisation was implemented by means of the Nelder–Mead method [2]. Results can be found in Tables 7.4–7.9.

Table 7.5 Relative errors in IP5

ϵ	$\lvert\frac{a_0^\epsilon - a_0}{a_0}\rvert$	$\lvert\frac{a_1^\epsilon - a_1}{a_1}\rvert$	$\lvert\frac{\alpha^\epsilon - \alpha}{\alpha}\rvert$	$\lvert\frac{\vartheta^\epsilon - \vartheta}{\vartheta}\rvert$	$\lvert\frac{\mu^\epsilon - \mu}{\mu}\rvert$	$\lvert\frac{\nu^\epsilon - \nu}{\nu}\rvert$
0.01%	0.06%	0.14%	0.006%	0.44%	0.05%	0.04%
0.1%	0.74%	1.9%	0.05%	7.7%	0.61%	0.34%
1%	9.6%	23%	0.33%	103%	7.2%	5.9%

Table 7.6 Relative errors in IP6

ϵ	$\lvert\frac{a_0^\epsilon - a_0}{a_0}\rvert$	$\lvert\frac{a_1^\epsilon - a_1}{a_1}\rvert$	$\lvert\frac{\alpha^\epsilon - \alpha}{\alpha}\rvert$	$\lvert\frac{\vartheta^\epsilon - \vartheta}{\vartheta}\rvert$	$\lvert\frac{\mu^\epsilon - \mu}{\mu}\rvert$
0.01%	0.04%	0.07%	0.003%	0.14%	0.02%
0.1%	0.55%	0.67%	0.03%	4.6%	0.22%
1%	6.6%	5.7%	0.11%	47%	3.4%

Table 7.7 Relative errors in IP7

ϵ	$\lvert\frac{a_0^\epsilon - a_0}{a_0}\rvert$	$\lvert\frac{a_1^\epsilon - a_1}{a_1}\rvert$	$\lvert\frac{\alpha^\epsilon - \alpha}{\alpha}\rvert$	$\lvert\frac{\vartheta^\epsilon - \vartheta}{\vartheta}\rvert$	$\lvert\frac{\mu^\epsilon - \mu}{\mu}\rvert$	$\lvert\frac{\nu^\epsilon - \nu}{\nu}\rvert$
0.01%	0.05%	0.09%	0.005%	0.38%	0.04%	0.03%
0.1%	0.65%	0.89%	0.04%	5.8%	0.55%	0.27%
1%	7.2%	15.0%	0.26%	79%	5.5%	2.6%

Table 7.8 Relative errors in IP8

ϵ	$\lvert\frac{a_1^\epsilon - a_1}{a_1}\rvert$	$\lvert\frac{\alpha^\epsilon - \alpha}{\alpha}\rvert$	$\lvert\frac{\vartheta^\epsilon - \vartheta}{\vartheta}\rvert$
0.01%	0.005%	0.002%	0.002%
0.1%	0.053%	0.014%	0.016%
1%	0.41%	0.09%	0.11%

Table 7.9 Relative errors in IP9

ϵ	$\lvert\frac{a_1^\epsilon - a_1}{a_1}\rvert$	$\lvert\frac{\alpha^\epsilon - \alpha}{\alpha}\rvert$	$\lvert\frac{\vartheta^\epsilon - \vartheta}{\vartheta}\rvert$	$\lvert\frac{\nu^\epsilon - \nu}{\nu}\rvert$
0.01%	0.005%	0.003%	0.003%	0.03%
0.1%	0.062%	0.022%	0.026%	0.19%
1%	0.61%	0.10%	0.14%	0.96%

Let us compare the results for IP1, IP2, IP4–IP7 with the corresponding results obtained for the inverse problems in linear models (Sect. 5.4.2). Tables 5.1, 5.3, 5.4, 7.1 and 7.2 show that the nonlinear waves are more informative concerning b and β but less informative concerning γ. Similarly, from Tables 5.2, 5.5 and 7.4, 7.5, 7.6, 7.7 we see that nonlinear waves give results that are better in α but worse in a_1. For other parameters of the coupled system the difference of results in nonlinear and linear cases is not remarkable.

Further, one can see from Tables 7.8 and 7.9 that the problems IP8 and IP9 give very good results, especially for ϑ. However, this occurs only in case the prescribed parameters a_0 and μ are exact. A noise in these parameters considerably worsens the results. For instance, if we perturb in IP9 the parameters a_0 and μ by errors given in columns 2 and 6 of Table 7.5 we obtain results that are not better than in IP5.

7.4 Proofs of Mathematical Statements

7.4.1 Proofs of Sect. 7.1.2

Counter-example for the uniqueness of the solution of IP2. To construct the counter-example, we need some auxiliary relations. For any two given triplets $S_i = (\beta_i, \gamma_i, \lambda_i)$, $i = 1, 2$, we define

$$\kappa_{ij} = \sqrt{\frac{c_j^2 - b}{\delta(\beta_i c_j^2 - \gamma_i)}}, \qquad \Theta_{ij} = -2\left[\frac{c_j^2 - b}{\beta_i c_j^2 - \gamma_i}\right]^{3/2}\frac{\lambda_i}{\mu}, \qquad i, j = 1, 2.$$

In an analogous way, the quantities κ_{ij}^0 and Θ_{ij}^0 with $i, j = 1, 2$, are defined via $S_i^0 = (\beta_i^0, \gamma_i^0, \lambda_i^0)$, $i = 1, 2$. Further, due to (6.14)

$$\xi^{\pm}[S_1, c_j](w) - \xi^{\pm}[S_2, c_j](w) \sim \mp\left(\frac{1}{\kappa_{1j}} - \frac{1}{\kappa_{2j}}\right)\ln|w| \quad \text{as } w \to 0 \tag{7.53}$$

and from (6.39) we deduce the relation

$$\xi^{\pm}[S_1, c_j](w) - \xi^{\pm}[S_2, c_j](w)$$
$$\sim -\frac{\sqrt{|w - A|}}{|A|}\left[\mp 2\left(\frac{1}{\kappa_{1j}} - \frac{1}{\kappa_{2j}}\right) + \frac{1}{2}\left(\frac{\Theta_{1j}}{\kappa_{1j}} - \frac{\Theta_{2j}}{\kappa_{2j}}\right)\sqrt{|w - A|}\right] \quad \text{as } w \to A. \tag{7.54}$$

Let us consider IP2 in case when the points P_{1l}, $l = 1, 2$, of the first wave are located at a common side of the extremum. This problem can be written in the form of the following system of nonlinear equations:

$$\xi^{\sigma_1}[S, c_1](w_{11}) = \xi_{11}, \qquad \xi^{\sigma_1}[S, c_1](w_{12}) = \xi_{12}, \qquad \xi^{\sigma_2}[S, c_2](w_2) = \xi_2 \tag{7.55}$$

where $\sigma_j \in \{+; -\}$, $j = 1, 2$. Let us choose some $S_1 = (\beta_1, \gamma_1, \lambda_1)$ and $S_2^0 = (\beta_2^0, \gamma_2^0, \lambda_2)$ so that

$$\beta_2^0 = \beta_1, \qquad \gamma_2^0 = \gamma_1 \quad \text{and} \quad \frac{\lambda_2}{\mu} > \frac{\lambda_1}{\mu}.$$

Then $\kappa_{1j} = \kappa^0_{2j}$ and from (7.54) we have

$$\xi^{\sigma_j}[S_1, c_j](w) - \xi^{\sigma_j}\big[S^0_2, c_j\big](w) \sim -\frac{|w-A|}{2|A|}\frac{1}{\kappa_{1j}}\big(\Theta_{1j} - \Theta^0_{2j}\big) \quad \text{as } w \to A$$

with $j = 1, 2$. Thus, since $\Theta_{1j} > \Theta^0_{2j}$, there exists $\epsilon > 0$ such that

$$\xi^{\sigma_j}[S_1, c_j](\omega_{2j}) - \xi^{\sigma_j}\big[S^0_2, c_j\big](\omega_{2j}) < 0, \quad j = 1, 2, \tag{7.56}$$

for ω_{2j} satisfying $|\omega_{2j} - A_j| = \epsilon$. The functions $\xi^\pm$ are analytical with respect to the parameters κ and Θ. (This follows from the representation of $\xi^\pm$ in the form of power series (6.38).) Consequently, (7.56) remains valid for small changes of S_2, i.e., there exists $\eta > 0$ such that

$$\xi^{\sigma_j}[S_1, c_j](\omega_{2j}) - \xi^{\sigma_j}[S_2, c_j](\omega_{2j}) < 0, \quad j = 1, 2, \tag{7.57}$$

if $|S_2 - S^0_2| < \eta$. Further, analysing the formula for κ_{ij} it is not difficult to see that it is possible to choose β_2 and γ_2 such that

$$\beta^2 \neq \beta^1, \qquad \gamma^2 \neq \gamma^1 \quad \text{and} \quad \sigma_j(\kappa_{2j} - \kappa_{1j}) < 0, \quad j = 1, 2, \tag{7.58}$$

and $S_2 - S^0_2 = (\beta_2 - \beta_1, \gamma_2 - \gamma_1, 0)$ is small enough, i.e., $|S_2 - S^0_2| < \eta$. In view of the latter inequality (7.57) holds and due to the latter property in (7.58) and the relations (7.53), (7.54) there exists $\epsilon_1 < \epsilon$ such that

$$\begin{aligned} &\xi^{\sigma_j}[S_1, c_j](\omega_{1j}) - \xi^{\sigma_j}[S_2, c_j](\omega_{1j}) > 0, \quad j = 1, 2,\\ &\xi^{\sigma_j}[S_1, c_j](\omega_{3j}) - \xi^{\sigma_j}[S_2, c_j](\omega_{3j}) > 0, \quad j = 1, 2, \end{aligned} \tag{7.59}$$

for ω_{1j} and ω_{3j} satisfying $|\omega_{1j}| = \epsilon_1$ and $|\omega_{3j} - A_j| = \epsilon_1$. Relations (7.57) and (7.59) imply that there exist $w_{11} \in (\omega_{11}, \omega_{12}), w_{12} \in (\omega_{12}, \omega_{13})$ and $w_2 \in (0, A_2)$ such that

$$\begin{aligned} &\xi^{\sigma_1}[S_1, c_1](w_{1l}) - \xi^{\sigma_1}[S_2, c_1](w_{1l}) = 0, \quad l = 1, 2,\\ &\xi^{\sigma_1}[S_1, c_2](w_2) - \xi^{\sigma_1}[S_2, c_2](w_2) = 0. \end{aligned}$$

Consequently, (7.55) or, equivalently, IP2 has two solutions $S_1 \neq S_2$ for such values of $w_{11} \neq w_{12}$ and w_2. □

Proof of Theorem 7.2 The inverse problem system reads either

$$\xi^+[S, c_1](w_{11}) = \xi_{11}, \qquad \xi^+[S, c_1](w_{12}) = \xi_{12}, \qquad \xi[S, c_2](w_2) = \xi_2 \tag{7.60}$$

when $P_{1l}, l = 1, 2$, are on the front side or

$$\xi^-[S, c_1](w_{11}) = \xi_{11}, \qquad \xi^-[S, c_1](w_{12}) = \xi_{12}, \qquad \xi[S, c_2](w_2) = \xi_2 \tag{7.61}$$

when $P_{1l}, l = 1, 2$, are on the rear side. Let us consider only the case (7.60) (the case (7.61) can be studied in a very similar manner). Note that $w_{11} \neq w_{12}$, because P_{1l},

$l = 1, 2$, are different. Without restriction of generality we may assume that w_{11} is closer to A_1 than w_{12}, i.e., $w_{11} \in (w_{12}, A_1)$. Supposing that (7.60) has two solutions $S = (\beta, \gamma, \lambda)$ and $\widetilde{S} = (\widetilde{\beta}, \widetilde{\gamma}, \widetilde{\lambda})$ we deduce from (7.60) the relations (7.6), (7.8) and

$$\xi^+[S, c_1](w_{12}) - \xi^+[\widetilde{S}, c_1](w_{12}) = 0. \tag{7.62}$$

Comparing the pairs (7.6) & (7.9), (7.6) & (7.62), (7.8) & (7.11) and applying Rolle's theorem we conclude that there exist $\overline{w}_{11} \in (w_{11}, A_1)$, $\overline{w}_{12} \in (w_{11}, w_{12})$ and $\overline{w}_2 \in (w_2, A_2)$ such that the equations

$$\xi^+[S, c_1]'(\overline{w}_{1l}) = \xi^+[\widetilde{S}, c_1]'(\overline{w}_{1l}) =: \xi'_{1l}, \quad l = 1, 2,$$
$$\xi[S, c_2]'(\overline{w}_2) = \xi[\widetilde{S}, c_2]'(\overline{w}_2) =: \xi'_2$$

hold. As in the proof of Theorem 7.1, this yields the homogeneous system (7.15) with the determinant (7.16). The determinant is different from zero because $c_1^2 \neq c_2^2$, $\xi'_{1l} \neq 0$ $(l = 1, 2)$, $\xi'_2 \neq 0$ and ξ'_{11}, ξ'_{12} are *different numbers* (this time they have a common sign). The difference of ξ'_{11} and ξ'_{12} immediately follows from the strict monotonicity of the derivative of $\xi^+[S, c_1](w)$ on the subinterval $(\frac{2A_1}{3}, A_1)$ (see Theorem 6.3) and the difference of the points $\overline{w}_{1l}$ in $(\frac{2A_1}{3}, A_1)$. Consequently, the solution of (7.15) is trivial, i.e., $\widetilde{S} = S$. The theorem is proved. □

Proof of Theorem 7.3 In a certain sense, IP3 can be interpreted as an extension of IP2. Therefore, the uniqueness in the cases when P_{1l} are located on both sides of the extremum automatically follows from Theorem 7.1. The uniqueness remains open only in the case when all points P_{1l}, $l = 1, 2, 3$, are located on a common side of the extremum. Let us study this case assuming that the location area of P_{1l}, $l = 1, 2, 3$, is the front side. (The study of the case of the rear side is similar.) Then IP3 is equivalent to the system

$$\xi^+[S, c_1](w_{1l}) = \xi_{1l}, \quad l = 1, 2, 3, \qquad \xi[S, c_2](w_2) = \xi_2.$$

Since P_{1l}, $l = 1, 2, 3$, are different, the quantities w_{1l}, $l = 1, 2, 3$, are also different. Therefore, we may assume without loss of generality that $w_{11} \in (w_{12}, A_1)$ and $w_{12} \in (w_{13}, A_1)$. Suppose that IP3 has two solutions $S = (\beta, \gamma, \lambda)$ and $\widetilde{S} = (\widetilde{\beta}, \widetilde{\gamma}, \widetilde{\lambda})$. Again, by means of Rolle's theorem, we deduce the following four relations:

$$\xi^+[S, c_1]'(\overline{w}_{1l}) = \xi^+[\widetilde{S}, c_1]'(\overline{w}_{1l}) =: \xi'_{1l}, \quad l = 1, 2, 3,$$
$$\xi[S, c_2]'(\overline{w}_2) = \xi[\widetilde{S}, c_2]'(\overline{w}_2) =: \xi'_2$$

with some points $\overline{w}_{1l}$, $l = 1, 2, 3$, such that $\overline{w}_{11} \in (w_{11}, A_1)$, $\overline{w}_{12} \in (w_{11}, w_{12})$, $\overline{w}_{13} \in (w_{12}, w_{13})$ and $\overline{w}_2 \in (w_2, A_2)$. They lead us to the over-determined homogeneous system

$$\begin{pmatrix} 3c_1^2\xi'_{11} & -3\xi'_{11} & 2\delta^{1/2} \\ 3c_1^2\xi'_{12} & -3\xi'_{12} & 2\delta^{1/2} \\ 3c_1^2\xi'_{13} & -3\xi'_{13} & 2\delta^{1/2} \\ 3c_2^2\xi'_2 & -3\xi'_2 & 2\delta^{1/2} \end{pmatrix} \begin{pmatrix} \widetilde{\beta} - \beta \\ \widetilde{\gamma} - \gamma \\ \widetilde{\lambda} - \lambda \end{pmatrix} = \begin{pmatrix} 0 \\ 0 \\ 0 \\ 0 \end{pmatrix} \tag{7.63}$$

instead of (7.15). The rank of the matrix of this system equals 3 because $c_1^2 \neq c_2^2$, $\xi'_{1l} \neq 0$ $(l = 1, 2, 3)$, $\xi'_2 \neq 0$ and the set $\{\xi'_{1l} : l = 1, 2, 3\}$ contains *at least two different elements*. The latter assertion is valid because the derivative of $\xi^+[S, c_1](w)$ has only two subintervals of strict monotonicity on $(0, A_1)$ (cf. Theorem 6.3) and all the three points $\overline{w}_{1l}$, $l = 1, 2, 3$, are different from each other. Consequently, the system (7.63) has only the trivial solution. This proves that $\widetilde{S} = S$. □

7.4.2 Proof of Theorem 7.5

We split this quite complicated proof into lemmas. The first lemma provides an auxiliary estimate for the inverses $\xi(w) = \xi^{\pm}(w)$ of the solitary wave solution with given parameters $\beta, \gamma, \lambda, b, \mu, c$. Recall that the amplitude equals $A = 3(c^2 - b)/\mu$.

Lemma 7.1 *For any $w, \overline{w}, \widehat{w} \in (0, A)$ the following estimates are valid:*

$$\frac{1}{\sqrt{5}}\left|\frac{\xi^{\pm'}(w)}{I_0'(w)}\right| < \left|\frac{\xi^{\pm}(\overline{w})}{I_0(\overline{w})}\right| < \sqrt{5}\left|\frac{\xi^{\pm'}(\widehat{w})}{I_0'(\widehat{w})}\right|. \tag{7.64}$$

Proof Dividing (6.32) by $\kappa^2\xi'(w)$ and taking the definition of I_0 in (6.37) into account we get the relation

$$\left(\frac{\xi'(w)}{I_0'(w)}\right)^2 = \frac{1}{\kappa^2}\left(1 - \frac{\Theta}{\kappa A\xi'(w)}\right).$$

Here $|\xi'(w)^{-1}| < \frac{2\kappa|A|}{3|\Theta|}$ according to the assertion (c) of Theorem 6.3. Thus, we deduce the inequalities

$$\frac{1}{3\kappa^2} < \left(\frac{\xi'(w)}{I_0'(w)}\right)^2 < \frac{5}{3\kappa^2} \tag{7.65}$$

for any $w \in (0, A)$. Using Cauchy's mean value theorem and the relations $\xi(A) = I_0(A) = 0$ we see that for any $w \in (0, A)$ there exists $v \in (w, A)$ such that $\frac{\xi'(v)}{I_0'(v)} = \frac{\xi(w)}{I_0(w)}$. Therefore, by means of the inequality (7.65) with w replaced by v we obtain the relation

$$\frac{1}{3\kappa^2} < \left(\frac{\xi(w)}{I_0(w)}\right)^2 < \frac{5}{3\kappa^2} \tag{7.66}$$

which holds for any $w \in (0, A)$. Combining (7.65) with (7.66) we deduce (7.64). The lemma is proved. □

In the sequel, let σ be the sign of the inverse function $\xi[S, c_2]$ in the nonlinear system (7.4) that is equivalent to IP2. In the next lemma we derive an estimate of the solution of $S = (\beta, \gamma, \lambda)$ in terms of the data d.

Lemma 7.2 *The following estimate holds*:

$$\max\{\delta|\beta|; \delta|\gamma|; \delta^{2/3}|\lambda|\} \leq \frac{2|\mu| \max\{1; c_1^2; c_2^2\} \max\{A_1^2; A_2^2\}}{3|c_1^2 - c_2^2|} K(d). \tag{7.67}$$

Proof Observing (7.4), the relations $\xi^{\pm}[S, c_j](A_j) = 0$, $j = 1, 2$, and applying Lagrange's mean value theorem we conclude that there exist $v_{1l} \in (w_{1l}, A_1)$, $l = 1, 2$ and $v_2 \in (w_2, A_2)$ such that

$$\begin{aligned} &\xi^+[S, c_1]'(v_{11}) = \frac{\xi_{11}}{w_{11} - A_1}, \qquad \xi^-[S, c_1]'(v_{12}) = \frac{\xi_{12}}{w_{12} - A_1}, \\ &\xi^\sigma[S, c_2]'(v_2) = \frac{\xi_2}{w_2 - A_2}. \end{aligned} \tag{7.68}$$

Further, let us write the equation (7.3) in the cases $\xi'(w) = \xi^+[S, c_1]'(w)$, $\xi'(w) = \xi^-[S, c_1]'(w)$ and $\xi'(w) = \xi^\sigma[S, c_2]'(w)$ with the arguments $w = v_{11}$, $w = v_{12}$ and $w = v_2$, respectively. Then, we get a 3×3 linear system

$$\mathcal{A} S_* = Y$$

for the vector $S_* = (\frac{3\delta}{\mu}\beta, \frac{3\delta}{\mu}\gamma, \frac{2\delta^{3/2}}{\mu}\lambda)^T$, where T stands for the transposition. Due to (7.68) the matrix and free term of this system read

$$\mathcal{A} = \begin{pmatrix} c_1^2 \xi_{11}(w_{11} - A_1)^{-1} & -\xi_{11}(w_{11} - A_1)^{-1} & 1 \\ c_1^2 \xi_{12}(w_{12} - A_1)^{-1} & -\xi_{12}(w_{12} - A_1)^{-1} & 1 \\ c_2^2 \xi_2(w_2 - A_2)^{-1} & -\xi_2(w_2 - A_2)^{-1} & 1 \end{pmatrix} \tag{7.69}$$

and

$$\begin{aligned} Y = \Bigg(&v_{11}^2(A_1 - v_{11}) \left[\frac{\xi_{11}}{w_{11} - A_1}\right]^3, \ v_{11}^2(A_1 - v_{11}) \left[\frac{\xi_{11}}{w_{11} - A_1}\right]^3, \\ &v_2^2(A_2 - v_2) \left[\frac{\xi_2}{w_2 - A_2}\right]^3 \Bigg)^T, \end{aligned} \tag{7.70}$$

respectively. Let us compute: $\det \mathcal{A} = \frac{(c_1^2 - c_2^2)\xi_2}{w_2 - A_2}[\frac{\xi_{11}}{w_{11} - A_1} - \frac{\xi_{12}}{w_{12} - A_1}]$. It holds $\operatorname{sign} \xi_{11} = -\operatorname{sign} \xi_{12}$, by the definition of ξ_{11}, ξ_{12}, and $\operatorname{sign}(w_{11} - A_1) =$

$\operatorname{sign}(w_{12} - A_1) = -\operatorname{sign} A_1$. Using these relations we obtain

$$|\det \mathcal{A}| = \frac{|c_1^2 - c_2^2||\xi_2|}{|w_2 - A_2|}\left[\frac{|\xi_{11}|}{|w_{11} - A_1|} + \frac{|\xi_{12}|}{|w_{12} - A_1|}\right].$$

By means of this formula, from (7.69) we deduce the following estimates for the components of the inverse matrix $\mathcal{A}^{-1} = (\widehat{a}_{ij})_{i,j=1,2,3}$:

$$|\widehat{a}_{ij}| \leq \frac{\max\{1; c_1^2; c_2^2\}(\frac{|\xi_{1i_*}|}{|w_{1i_*} - A_1|} + \frac{|\xi_2|}{|w_2 - A_2|})}{|\det \mathcal{A}|}$$

$$\leq \frac{\max\{1; c_1^2; c_2^2\}}{|c_1^2 - c_2^2|}\left[\frac{|w_{11} - A_1|}{|\xi_{11}|} + \frac{|w_{12} - A_1|}{|\xi_{12}|} + \frac{|w_2 - A_2|}{|\xi_2|}\right], \quad i, j = 1, 2,$$

$$|\widehat{a}_{3j}| \leq \frac{\max\{1; c_1^2\}(\frac{|\xi_{11}|}{|w_{11} - A_1|} + \frac{|\xi_{12}|}{|w_{12} - A_1|})}{|\det \mathcal{A}|} \leq \frac{\max\{1; c_1^2\}}{|c_1^2 - c_2^2|}\frac{|w_2 - A_2|}{|\xi_2|}, \quad j = 1, 2,$$

$$|\widehat{a}_{i3}| \leq \frac{|c_1^2 - c_2^2|\frac{|\xi_{1i_*}|}{|w_{1i_*} - A_1|}\frac{|\xi_2|}{|w_2 - A_2|}}{|\det \mathcal{A}|} \leq 1, \quad i = 1, 2$$

and $\widehat{a}_{33} = 0$. Here $i_* = 1$ if $i = 2$ and $i_* = 2$ if $i = 1$. Summing up,

$$|\widehat{a}_{ij}| \leq \frac{2\max\{1; c_1^2; c_2^2\}}{|c_1^2 - c_2^2|}\left[1 + \frac{|w_{11} - A_1|}{|\xi_{11}|} + \frac{|w_{12} - A_1|}{|\xi_{12}|} + \frac{|w_2 - A_2|}{|\xi_2|}\right] \tag{7.71}$$

for $i, j = 1, 2, 3$. Further, some terms in (7.70) are estimated as follows:

$$\begin{aligned} &|v_{1l}| \leq |A_1|, \qquad |A_1 - v_{1l}| \leq |A_1 - w_{1l}|, \quad l = 1, 2, \\ &|v_2| \leq |A_2|, \qquad |A_2 - v_2| \leq |A_2 - w_2|. \end{aligned} \tag{7.72}$$

Using (7.70)–(7.72) in the equation $S_* = \mathcal{A}^{-1}Y$ we prove (7.67) with (7.25). The proof of the lemma is complete. □

Furthermore, we prove an additional technical lemma for the inverses of the solitary wave solution $\xi^{\pm}[S^i]$ that correspond to two triplets $S^i = (\beta^i, \gamma^i, \lambda^i)$, $i = 1, 2$, and the given parameters b, μ, c.

Lemma 7.3 *For any $w^i \in (0, A)$, $i = 1, 2$, there exist $u_i^{\pm} = u_i^{\pm}(w^1, w^2, A) \in (w^i, A)$, $i = 1, 2$, such that the estimates*

$$\begin{aligned} &|\xi^{\pm}[S^1]'(u_1^{\pm}) - \xi^{\pm}[S^2]'(u_2^{\pm})| \\ &\quad \leq \frac{C_1^*}{r}\left[|\xi^{\pm}[S^1](w^1) - \xi^{\pm}[S^2](w^2)| + \frac{M^{\pm}}{|w^2|r^{1/2}}|w^1 - w^2|\right] \end{aligned} \tag{7.73}$$

and

$$\begin{aligned}&\left|(u_1^{\pm})^2(A-u_1^{\pm})(\xi^{\pm}[S^1]'(u_1^{\pm}))^3-(u_2^{\pm})^2(A-u_2^{\pm})(\xi^{\pm}[S^2]'(u_2^{\pm}))^3\right|\\&\quad\le\frac{C_2^*(M^{\pm})^2}{r}\left[\left|\xi^{\pm}[S^1](w^1)-\xi^{\pm}[S^2](w^2)\right|+\frac{M^{\pm}}{q^3r^{1/2}}|w^1-w^2|\right]\end{aligned}\tag{7.74}$$

hold. Here

$$M^{\pm}=\max_{i=1,2}\frac{|\xi^{\pm}[S^i](w^i)|}{|I_0(w^i)|},\qquad q=\min_{i=1,2}\left|w^i\right|,\qquad r=\min_{i=1,2}\left|w^i-A\right|$$

and C_1^*, C_2^* are some constants depending on $|A|$.

Proof Let us define the functions

$$g^{\pm}(t)=\xi^{\pm}[S^1]\big(m^1(t)\big)-\xi^{\pm}[S^2]\big(m^2(t)\big)$$

for $t\in[0,1]$ where

$$m^i(t)=A+t\big(w^i-A\big)-2\,\mathrm{sign}\,Ar\big(t^{2/3}-t\big),\quad t\in[0,1],\ i=1,2.\tag{7.75}$$

One can immediately check that the functions m^i, $i=1,2$, are strictly monotonic, $m^i(0)=A$, $m^i(1)=w^i$, $i=1,2$, and the following relations are valid:

$$\begin{aligned}&\left|(m^i)'(t)\right|\ge\frac{r}{3},\quad i=1,2,\qquad\big(m^2-m^1\big)'(t)=w^2-w^1,\\&\left|\sqrt{1-\frac{m^2(t)}{A}}\big(m^1\big)'(t)\right|\ge\frac{r^{3/2}}{3\sqrt{|A|}},\end{aligned}\tag{7.76}$$

$$\left|m^i(t)\right|\le|A|,\qquad\left|m^i(t)\right|\ge\left|w^i\right|,\qquad\left|A-m^i(t)\right|\ge rt^{2/3},\quad i=1,2.\tag{7.77}$$

Observing that $g^{\pm}(0)=\xi^{\pm}[S^1](A)-\xi^{\pm}[S^2](A)=0$ and using Lagrange's mean value theorem we see that there exist $\tau^{\pm}\in(0,1)$ such that

$$g^{\pm'}\big(\tau^{\pm}\big)=g^{\pm}(1).$$

Let us denote

$$u_i^{\pm}=m^i\big(\tau^{\pm}\big)\in\big(w^i,A\big),\quad i=1,2.\tag{7.78}$$

Remark that $u_i^{\pm}$ depend on w^1, w^2 and A.

For sake of simplicity, let us drop the superscript $\pm$ in the rest of the proof. Due to the definition of g, the relation $g'(\tau)=g(1)$ has the form

$$\big(m^1\big)'(\tau)\xi\big[S^1\big]'(u_1)-\big(m^2\big)'(\tau)\xi\big[S^2\big]'(u_2)=\xi\big[S^1\big]\big(w^1\big)-\xi\big[S^2\big]\big(w^2\big).$$

This implies that

$$\xi\big[S^1\big]'(u_1) - \xi\big[S^2\big]'(u_2) = \frac{\xi[S^1](w^1) - \xi[S^2](w^2)}{(m^1)'(\tau)} + \frac{\xi[S^2]'(u_2)(m^2 - m^1)'(\tau)}{(m^1)'(\tau)}. \tag{7.79}$$

By means of Lemma 7.1, the formula $I_0'(w) = [w\sqrt{1-\frac{w}{A}}]^{-1}$ and (7.76), (7.77) we deduce that

$$\begin{aligned} \left|\frac{\xi[S^2]'(u_2)}{(m^1)'(\tau)}\right| &< \left|\frac{\sqrt{5}I_0'(u_2)\xi[S^2](w^2)}{(m^1)'(\tau)I_0(w^2)}\right| \\ &= \left|m^2(\tau)\sqrt{1-\frac{m^2(\tau)}{A}}\big(m^1\big)'(\tau)\right|^{-1}\frac{\sqrt{5}|\xi[S^2](w^2)|}{|I_0(w^2)|} \\ &\le \frac{3\sqrt{5|A|}M}{|w^2|r^{3/2}}. \end{aligned} \tag{7.80}$$

From (7.79), in view of (7.76) and (7.80), we obtain (7.73).

Next let us prove the relation (7.74). Using the formula for $I_0'(w)$ above we obtain

$$\begin{aligned} &u_1^2(A-u_1)\big(\xi\big[S^1\big]'(u_1)\big)^3 - u_2^2(A-u_2)\big(\xi\big[S^2\big]'(u_2)\big)^3 \\ &= A\bigg[\bigg(\frac{\xi[S^1]'(u_1)}{I_0'(u_1)}\bigg)^2 + \bigg(\frac{\xi[S^2]'(u_2)}{I_0'(u_2)}\bigg)^2 + \frac{\xi[S^1]'(u_1)\xi[S^2]'(u_2)}{I_0'(u_1)I_0'(u_2)}\bigg] \\ &\times \big(\xi\big[S^1\big]'(u_1) - \xi\big[S^2\big]'(u_2)\big) \\ &+ |A|^{3/2}\frac{\xi[S^1]'(u_1)\xi[S^2]'(u_2)}{I_0'(u_1)I_0'(u_2)}\bigg[\frac{\xi[S^1]'(u_1)}{I_0'(u_1)} + \frac{\xi[S^2]'(u_2)}{I_0'(u_2)}\bigg] \\ &\times \frac{u_1^2(A-u_1) - u_2^2(A-u_2)}{\prod_{i=1,2} u_i\sqrt{|A-u_i|}[u_1\sqrt{|A-u_1|} + u_2\sqrt{|A-u_2|}]}. \end{aligned} \tag{7.81}$$

To estimate this relation, we derive some auxiliary inequalities. By (7.77) we have

$$|u_i|\sqrt{|A-u_i|} = \big|m^i(\tau)\big|\sqrt{\big|A-m^i(\tau)\big|} \ge \big|w^i\big|r^{1/2}\tau^{1/3} \ge qr^{1/2}\tau^{1/3}.$$

Moreover, using again (7.77) and the relation $(m^1 - m^2)(\tau) = (w^1 - w^2)\tau$ following from (7.75), we compute

$$\begin{aligned} &\big|u_1^2(A-u_1) - u_2^2(A-u_2)\big| \\ &= \big|\big[A\big(m^1 + m^2\big)(\tau) - \big(\big(m^1\big)^2 + m^1m^2 + \big(m^2\big)^2\big)(\tau)\big]\big(m^1 - m^2\big)(\tau)\big| \\ &\le 5|A|^2\big|w^1 - w^2\big|\tau. \end{aligned}$$

Thus,

$$\left|\frac{u_1^2(A-u_1)-u_2^2(A-u_2)}{\prod_{i=1,2}u_i\sqrt{|A-u_i|}[u_1\sqrt{|A-u_1|}+u_2\sqrt{|A-u_2|}]}\right| \le \frac{5|A|^2|w^1-w^2|}{2q^3r^{3/2}}. \tag{7.82}$$

Finally, using the proved estimate (7.73) (there we take $|w^2| \ge \frac{q^3}{|A|^2}$), the relation (7.82) and Lemma 7.1 in the formula (7.81), we deduce the desired estimate (7.74). The lemma is proved. □

Now we are able to complete the proof of Theorem 7.5. For our convenience we re-denote the exact and approximate data and the related solutions of IP2 by means of superscripts, namely

$$d^1=(w_{11}^1,w_{12}^1,w_2^1,\xi_{11}^1,\xi_{12}^1,\xi_2^1)=d=(w_{11},w_{12},w_2,\xi_{11},\xi_{12},\xi_2),$$
$$d^2=(w_{11}^2,w_{12}^2,w_2^2,\xi_{11}^2,\xi_{12}^2,\xi_2^2)=\widetilde{d}=(\widetilde{w}_{11},\widetilde{w}_{12},\widetilde{w}_2,\widetilde{\xi}_{11},\widetilde{\xi}_{12},\widetilde{\xi}_2),$$
$$S^1=(\beta^1,\gamma^1,\lambda^1)=S=(\beta,\gamma,\lambda), \qquad S^2=(\beta^2,\gamma^2,\lambda^2)=\widetilde{S}=(\widetilde{\beta},\widetilde{\gamma},\widetilde{\lambda}).$$

Moreover, let $\sigma \in \{+;-\}$ be the upper sign of the function $\xi[S,c_2]$ in the last equation of the inverse problem system (7.4).

Using the functions $u_i^{\pm}$ from Lemma 7.3 we introduce the following quantities:

$$u_{11}^i=u_i^+(w_{11}^1,w_{11}^2,A_1), \qquad u_{12}^i=u_i^-(w_{12}^1,w_{12}^2,A_1),$$
$$u_2^i=u_i^\sigma(w_2^1,w_2^2,A_2), \quad i=1,2,$$

$$\psi_{11}^i=\xi^+[S^i,c_1]'(u_{11}^i), \qquad \psi_{12}^i=\xi^-[S^i,c_1]'(u_{12}^i),$$
$$\psi_2^i=\xi^\pm[S^i,c_2]'(u_2^i), \quad i=1,2. \tag{7.83}$$

We write (7.3) for $\xi'(w)=\xi^+[S^i,c_1]'(w)$, $\xi'(w)=\xi^-[S^i,c_1]'(w)$ and $\xi'(w)=\xi^\sigma[S^i,c_2]'(w)$, with $w=u_{11}^i$, $w=u_{12}^i$ and $w=u_2^i$, respectively, where $i=1,2$. This yields 3×3 systems

$$\mathcal{B}_i S_*^i = Z^i, \quad i=1,2,$$

where

$$S_*^i=\left(\frac{3\delta}{\mu}\beta^i,\frac{3\delta}{\mu}\gamma^i,\frac{2\delta^{3/2}}{\mu}\lambda^i\right)^T, \quad i=1,2.$$

In view of (7.83) the matrices and free terms of these systems are representable as follows:

$$\mathcal{B}_i = \begin{pmatrix} c_1^2\psi_{11}^i & -\psi_{11}^i & 1 \\ c_1^2\psi_{12}^i & -\psi_{12}^i & 1 \\ c_2^2\psi_2^i & -\psi_2^i & 1 \end{pmatrix}, \quad i = 1, 2,$$

$$Z^i = \big((u_{11}^i)^2(A_1 - u_{11}^i)(\psi_{11}^i)^3,\ (u_{12}^i)^2(A_1 - u_{12}^i)(\psi_{12}^i)^3,$$
$$(u_2^i)^2(A_2 - u_2^i)(\psi_2^i)^3\big)^T, \quad i = 1, 2.$$

The difference $S_*^1 - S_*^2$ can be expressed in the form

$$S_*^1 - S_*^2 = (\mathcal{B}_1)^{-1}V$$

where

$$V = Z^1 - Z^2 + (\mathcal{B}_2 - \mathcal{B}_1)S_*^2.$$

To prove the assertion (7.22), we have to estimate $(\mathcal{B}_1)^{-1}$ and V. Observe that $\det\mathcal{B}_1 = (c_1^2 - c_2^2)\psi_2^1[\psi_{11}^1 - \psi_{12}^1]$. According to the definition of ξ^+ and ξ^-, the numbers $\psi_{11}^1 = \xi^+[S^1, c_1]'(u_{11}^1)$ and $\psi_{12}^1 = \xi^-[S^1, c_1]'(u_{12}^1)$ have different signs. Thus, we get

$$|\det\mathcal{B}_1| = |c_1^2 - c_2^2||\psi_2^1|[|\psi_{11}^1| + |\psi_{12}^1|].$$

As in the proof of Lemma 7.2, we derive the following estimate for the components of the inverse $\mathcal{B}_1^{-1} = (\widehat{b}_{ij})_{i,j=1,2,3}$:

$$|\widehat{b}_{ij}| \le \frac{2\max\{1; c_1^2; c_2^2\}}{|c_1^2 - c_2^2|}\left[1 + \frac{1}{|\psi_{11}^1|} + \frac{1}{|\psi_{12}^1|} + \frac{1}{|\psi_2^1|}\right], \quad i, j = 1, 2, 3. \tag{7.84}$$

Using (7.64), the definitions of ψ_{11}^1, ξ_{11}^1, $I_0[A](w)$ and the inequalities $|w_{11}^1| < |u_{11}^1| < |A_1|$ we obtain

$$|\psi_{11}^1| > \frac{|I_0[A_1]'(u_{11}^1)||\xi_{11}^1|}{\sqrt{5}|I_0[A_1](w_{11}^1)|} = \frac{[u_{11}^1\sqrt{1 - \frac{u_{11}^1}{A_1}}]^{-1}|\xi_{11}|}{\sqrt{5}|I_0[A_1](w_{11}^1)|}$$
$$> \frac{[\sqrt{|A_1 - w_{11}|}]^{-1}|\xi_{11}|}{\sqrt{5}|I_0[A_1](w_{11}^1)|} \ge \frac{1}{\sqrt{5}L(d)}.$$

Similarly we get $|\psi_{12}^1| > \frac{1}{\sqrt{5}L(d)}$ and $|\psi_2^1| > \frac{1}{\sqrt{5}L(d)}$. Using these relations in (7.84) we deduce that

$$|\widehat{b}_{ij}| \le \frac{6\sqrt{5}\max\{1; c_1^2; c_2^2\}}{|c_1^2 - c_2^2|}(1 + L(d)), \quad i, j = 1, 2, 3. \tag{7.85}$$

Next let us estimate the vector $V = (V_1, V_2, V_3)^T = Z^1 - Z^2 + (\mathcal{B}_2 - \mathcal{B}_1)S_*^2$. For the first component

$$V_1 = (u_{11}^1)^2(A_1 - u_{11}^1)(\psi_{11}^1)^3 - (u_{11}^2)^2(A_1 - u_{11}^2)(\psi_{11}^2)^3$$
$$+ c_1^2(\psi_{11}^2 - \psi_{11}^1)\frac{3\delta}{\mu}\beta^2 - (\psi_{11}^2 - \psi_{11}^1)\frac{3\delta}{\mu}\gamma^2,$$

by virtue of Lemmas 7.2 and 7.3, we get

$$|V_1| \le \frac{C_2^*(M^+)^2}{r}\left[|\xi_{11}^1 - \xi_{11}^2| + \frac{M^+}{q^3 r^{1/2}}|w_{11}^1 - w_{11}^2|\right]$$
$$+ \frac{C_3^* K(d^2)}{|c_1^2 - c_2^2|r}\left[|\xi_{11}^1 - \xi_{11}^2| + \frac{M^+}{q r^{1/2}}|w_{11}^1 - w_{11}^2|\right]$$
$$\le \frac{C_4^*}{|c_1^2 - c_2^2|}\left[\frac{[Q(d^1,d^2)]^2 + K(d^2)}{z_A(d^1,d^2)}\varepsilon_\xi\right.$$
$$\left. + \frac{Q(d^1,d^2)}{[z_A(d^1,d^2)]^{3/2}}\left(\frac{[Q(d^1,d^2)]^2}{[z_0(d^1,d^2)]^3} + \frac{K(d^2)}{z_0(d^1,d^2)}\right)\varepsilon_w\right]. \qquad (7.86)$$

In this relation

$$M^+ = \max_{i=1,2}\frac{|\xi_{11}^i|}{|I_0[A_1](w_{11}^i)|}, \qquad q = \min_{i=1,2}|w_{11}^i|, \qquad r = \min_{i=1,2}|w_{11}^i - A_1|$$

and C_3^*, C_4^* are constants depending on μ, $|A_1|$ and $\max_{i=1,2}|c_i|$. We obtain similar estimates for the other components, too:

$$|V_2|, |V_3| \le \frac{C_4^*}{|c_1^2 - c_2^2|}\left[\frac{[Q(d^1,d^2)]^2 + K(d^2)}{z_A(d^1,d^2)}\varepsilon_\xi\right.$$
$$\left. + \frac{Q(d^1,d^2)}{[z_A(d^1,d^2)]^{3/2}}\left(\frac{[Q(d^1,d^2)]^2}{[z_0(d^1,d^2)]^3} + \frac{K(d^2)}{z_0(d^1,d^2)}\right)\varepsilon_w\right]. \qquad (7.87)$$

Finally, estimating $S_*^1 - S_*^2 = (\mathcal{B}_1)^{-1}V$ by means of (7.85)–(7.87) we deduce (7.22). Theorem 7.5 is completely proved. □

7.4.3 Proofs of Sect. 7.2.2

Proof of Theorem 7.7 Firstly, let us order the set of points P_{il} in a suitable manner. Let the points P_{il}, $l = 1, \ldots, l_i$, $i = 1, 2$, stand on the front sides of the waves and P_{il}, $l = l_i + 1, \ldots, 16$, $i = 1, 2$, stand on the rear sides of the waves. In particular cases when either $l_i = 0$ or $l_i = 16$ for some $i \in \{1; 2\}$, all P_{il} are located only on the rear or front side, respectively. Moreover, let the levels be ordered away from the amplitude. This means that

$$w_{i1} \in (w_{i2}, A_i), \qquad w_{i2} \in (w_{i3}, A_i), \quad \ldots, \quad w_{i,l_i-1} \in (w_{il_i}, A_i),$$
$$w_{i,l_i+1} \in (w_{i,l_i+2}, A_i), \qquad w_{i,l_i+2} \in (w_{i,l_i+3}, A_i), \quad \ldots, \quad w_{i,15} \in (w_{i,16}, A_i)$$

for $i = 1, 2$. Then IP5 is equivalent to the following nonlinear system for the vector $S = (a_0, a_1, \vartheta, \alpha, \mu, \nu)$:

$$\begin{aligned} \xi^+[S, c_i](w_{il}) &= \xi_{il}, \quad l = 1, \ldots, l_i, \\ \xi^-[S, c_i](w_{il}) &= \xi_{il}, \quad l = l_i + 1, \ldots, 16, \; i = 1, 2, \end{aligned} \tag{7.88}$$

$$\xi^\pm[S, c_i](A_i) = 0, \quad i = 1, 2. \tag{7.89}$$

Suppose that IP5 has another solution $\widetilde{S} = (\widetilde{a}_0, \widetilde{a}_1, \widetilde{\vartheta}, \widetilde{\alpha}, \widetilde{\mu}, \widetilde{\nu})$, too. This means that

$$\begin{aligned} \xi^+[\widetilde{S}, c_i](w_{il}) &= \xi_{il}, \quad l = 1, \ldots, l_i, \\ \xi^-[\widetilde{S}, c_i](w_{il}) &= \xi_{il}, \quad l = l_i + 1, \ldots, 16, \; i = 1, 2, \end{aligned} \tag{7.90}$$

$$\xi^\pm[\widetilde{S}, c_i](A_i) = 0, \quad i = 1, 2. \tag{7.91}$$

Let us apply Rolle's theorem for the functions in the relations (7.88)–(7.91). This yields that there exist numbers

$$\begin{aligned} &\overline{w}_{i1} \in (w_{i1}, A_1), \qquad \overline{w}_{i2} \in (w_{i2}, w_{i1}), \quad \ldots, \quad \overline{w}_{il_i} \in (w_{il_i}, w_{i,l_i-1}), \\ &\overline{w}_{i,l_i+1} \in (w_{i,l_i+1}, A_1), \qquad \overline{w}_{i,l_i+2} \in (w_{i,l_i+2}, w_{i,l_i+1}), \quad \ldots, \\ &\overline{w}_{l,16} \in (w_{l,16}, w_{l,15}) \end{aligned} \tag{7.92}$$

for $i = 1, 2$ such that the equations

$$\begin{aligned} \xi^+[S, c_i]'(\overline{w}_{il}) &= \xi^+[\widetilde{S}, c_i]'(\overline{w}_{il}) = \xi'_{il}, \quad l = 1, \ldots, l_i, \\ \xi^-[S, c_i]'(\overline{w}_{il}) &= \xi^-[\widetilde{S}, c_i]'(\overline{w}_{il}) = \xi'_{il}, \quad l = l_i + 1, \ldots, 16, \; i = 1, 2 \end{aligned} \tag{7.93}$$

are valid with some numbers ξ'_{il}. Plugging the data from (7.93) into (7.31) we transform (7.93) to the pair of polynomial systems

$$\begin{aligned} &3\delta\left(c_i^2 - a_1\right)\left[c_i^2 - a_0 - \mu\overline{w}_{il}\right]^2 \xi'_{il} - 2\delta^{3/2}\nu\left[c_i^2 - a_0 - \mu\overline{w}_{il}\right]^3 \\ &= -\overline{w}_{il}^2 \left\{3\alpha\left[c_i^2 - a_0 - \frac{\mu}{2}\overline{w}_{il}\right]^2 + 3\vartheta\left(c_i^2 - a_0\right) - 2\vartheta\mu\overline{w}_{il}\right\}\left[\xi'_{il}\right]^3, \\ &l = 1, \ldots, 16, \; i = 1, 2, \end{aligned} \tag{7.94}$$

$$\begin{aligned} &3\delta\left(c_i^2 - \widetilde{a}_1\right)\left[c_i^2 - \widetilde{a}_0 - \widetilde{\mu}\overline{w}_{il}\right]^2 \xi'_{il} - 2\delta^{3/2}\widetilde{\nu}\left[c_i^2 - \widetilde{a}_0 - \widetilde{\mu}\overline{w}_{il}\right]^3 \\ &= -\overline{w}_{il}^2 \left\{3\widetilde{\alpha}\left[c_i^2 - \widetilde{a}_0 - \frac{\widetilde{\mu}}{2}\overline{w}_{il}\right]^2 + 3\widetilde{\vartheta}\left(c_i^2 - \widetilde{a}_0\right) - 2\widetilde{\vartheta}\widetilde{\mu}\overline{w}_{il}\right\}\left[\xi'_{il}\right]^3, \\ &l = 1, \ldots, 16, \; i = 1, 2. \end{aligned} \tag{7.95}$$

The degree of the system (7.94) is 4. This is the reason why the uniqueness in the balanced case cannot be expected and the number of equations was to be increased.

We will show that the chosen number $16+16=32$ of equations of the form (7.94) together with the 2 amplitude relations are sufficient to guarantee the uniqueness. In the sequel we can assume that either ν or $\widetilde{\nu}$ is different from zero, because the case $\nu=\widetilde{\nu}=0$ is subject to Theorem 7.6. More precisely, let $\nu \neq 0$. (When $\widetilde{\nu} \neq 0$ we can interchange S and $\widetilde{S}$.)

Let us eliminate ξ'_{il} from (7.94), (7.95). To simplify this procedure, we introduce the following polynomials:

$$
\begin{aligned}
&Q_{2,i}(w) = 3\delta\bigl(c_i^2 - a_1\bigr)\bigl[c_i^2 - a_0 - \mu w\bigr]^2,\\
&\widetilde{Q}_{2,i}(w) = 3\delta\bigl(c_i^2 - \widetilde{a}_1\bigr)\bigl[c_i^2 - \widetilde{a}_0 - \widetilde{\mu} w\bigr]^2,\\
&R_{3,i}(w) = 2\delta^{3/2}\nu\bigl[c_i^2 - a_0 - \mu w\bigr]^3, \qquad \widetilde{R}_{3,i}(w) = 2\delta^{3/2}\widetilde{\nu}\bigl[c_i^2 - \widetilde{a}_0 - \widetilde{\mu} w\bigr]^3,\\
&T_{2,i}(w) = 3\alpha\left[c_i^2 - a_0 - \frac{\mu}{2}w\right]^2 + 3\vartheta\bigl(c_i^2 - a_0\bigr) - 2\vartheta\mu w,\\
&\widetilde{T}_{2,i}(w) = 3\widetilde{\alpha}\left[c_i^2 - \widetilde{a}_0 - \frac{\widetilde{\mu}}{2}w\right]^2 + 3\widetilde{\vartheta}\bigl(c_i^2 - \widetilde{a}_0\bigr) - 2\widetilde{\vartheta}\widetilde{\mu} w
\end{aligned}
$$

for $i=1,2$. Multiplying (7.94) by $\widetilde{Q}_{2,i}(\overline{w}_{il})$, (7.95) by $Q_{2,i}(\overline{w}_{il})$ and subtracting, we deduce that

$$
\begin{aligned}
&\overline{w}_{il}^2\bigl\{\widetilde{Q}_{2,i}(\overline{w}_{il})T_{2,i}(\overline{w}_{il}) - Q_{2,i}(\overline{w}_{il})\widetilde{T}_{2,i}(\overline{w}_{il})\bigr\}\bigl[\xi'_{il}\bigr]^3\\
&\quad = R_{3,i}(\overline{w}_{il})\widetilde{Q}_{2,i}(\overline{w}_{il}) - \widetilde{R}_{3,i}(\overline{w}_{il})Q_{2,i}(\overline{w}_{il}),\\
&\quad l=1,\ldots,16,\ i=1,2.
\end{aligned}
\tag{7.96}
$$

Furthermore, multiplying (7.94) by $\widetilde{T}_{2,i}(\overline{w}_{il})$ and (7.95) by $T_{2,i}(\overline{w}_{il})$, subtracting and cubing the result, we deduce that

$$
\begin{aligned}
&\bigl\{Q_{2,i}(\overline{w}_{il})\widetilde{T}_{2,i}(\overline{w}_{il}) - \widetilde{Q}_{2,i}(\overline{w}_{il})T_{2,i}(\overline{w}_{il})\bigr\}^3\bigl[\xi'_{il}\bigr]^3\\
&\quad = \bigl\{R_{3,i}(\overline{w}_{il})\widetilde{T}_{2,i}(\overline{w}_{il}) - \widetilde{R}_{3,i}(\overline{w}_{il})T_{2,i}(\overline{w}_{il})\bigr\}^3,\\
&\quad l=1,\ldots,16,\ i=1,2.
\end{aligned}
\tag{7.97}
$$

Finally, multiplying (7.96) by $\{Q_{2,i}(\overline{w}_{il})\widetilde{T}_{2,i}(\overline{w}_{il}) - \widetilde{Q}_{2,i}(\overline{w}_{il})T_{2,i}(\overline{w}_{il})\}^2$ and (7.97) by $\overline{w}_{il}^2$ and adding the obtained equalities, the quantities ξ'_{il} are eliminated. We obtain

$$
\begin{aligned}
&\bigl\{R_{3,i}(\overline{w}_{il})\widetilde{Q}_{2,i}(\overline{w}_{il}) - \widetilde{R}_{3,i}(\overline{w}_{il})Q_{2,i}(\overline{w}_{il})\bigr\}\\
&\quad \times \bigl\{Q_{2,i}(\overline{w}_{il})\widetilde{T}_{2,i}(\overline{w}_{il}) - \widetilde{Q}_{2,i}(\overline{w}_{il})T_{2,i}(\overline{w}_{il})\bigr\}^2\\
&\quad - \overline{w}_{il}^2\bigl\{R_{3,i}(\overline{w}_{il})\widetilde{T}_{2,i}(\overline{w}_{il}) - \widetilde{R}_{3,i}(\overline{w}_{il})T_{2,i}(\overline{w}_{il})\bigr\}^3 = 0,\\
&l=1,\ldots,16,\ i=1,2.
\end{aligned}
$$

This shows that $\overline{w}_{il}, l = 1, \ldots, 16$, $i = 1, 2$, are the roots of the following polynomial of the 17th degree:

$$\mathcal{P}_{17,i}(w) = \{R_{3,i}(w)\widetilde{Q}_{2,i}(w) - \widetilde{R}_{3,i}(w)Q_{2,i}(w)\}$$
$$\times \{Q_{2,i}(w)\widetilde{T}_{2,i}(w) - \widetilde{Q}_{2,i}(w)T_{2,i}(w)\}^2 \tag{7.98}$$
$$- w^2\{R_{3,i}(w)\widetilde{T}_{2,i}(w) - \widetilde{R}_{3,i}(w)T_{2,i}(w)\}^3, \quad i = 1, 2. \tag{7.99}$$

It follows from (7.92) that any of the subsets of roots $\mathcal{W}_i^+ = \{\overline{w}_{il},\ l = 1, \ldots, l_i\}$, $\mathcal{W}_i^- = \{\overline{w}_{il},\ l = l_i + 1, \ldots, 16\}$, $i = 1, 2$, consists of different elements. However, these subsets may intersect. More precisely, for any $i \in \{1; 2\}$ it may happen that

$$\exists \overline{w}_{i,\lambda_1} \in \mathcal{W}_i^+, \overline{w}_{i,\lambda_2} \in \mathcal{W}_i^- : \quad \overline{w}_{i,\lambda_1} = \overline{w}_{i,\lambda_2}. \tag{7.100}$$

Let us suppose that for some $i \in \{1; 2\}$ the intersection is nonempty, i.e., (7.100) is valid. Setting $\lambda = \lambda_1$ and $\lambda = \lambda_2$ in (7.94), subtracting the obtained systems and using (7.100) we have

$$\overline{w}_{1,\lambda_1} T_{2,i}(\overline{w}_{1,\lambda_1})\{[\xi'_{i,\lambda_1}]^3 - [\xi'_{i,\lambda_2}]^3\} = 0. \tag{7.101}$$

Since ν is different from zero, the waves under consideration are asymmetric. This implies that $\xi'_{i,\lambda_1} \neq \xi'_{i,\lambda_2}$. Therefore, (7.101) with (7.100) and $\overline{w}_{1,\lambda_1} \neq 0$ yields $T_{2,i}(\overline{w}_{1,\lambda_1}) = T_{2,i}(\overline{w}_{1,\lambda_2}) = 0$. Similarly we derive $\widetilde{T}_{2,i}(\overline{w}_{1,\lambda_1}) = \widetilde{T}_{2,i}(\overline{w}_{1,\lambda_2}) = 0$. By virtue of these equalities and (7.98), the number $\overline{w}_{1,\lambda_1} = \overline{w}_{1,\lambda_2}$ is a double root of the polynomial $\mathcal{P}_{17,i}$. Now we come to a conclusion that, for both $i = 1, 2$, the set of numbers $\{\overline{w}_{il},\ l = 1, \ldots, 16\}$ consists of roots of $\mathcal{P}_{17,i}$ with total multiplicity 16. Furthermore, by (7.41) $T_{2,i}(A_i) = \widetilde{T}_{2,i}(A_i) = 0$, $i = 1, 2$. Therefore, for both $i = 1, 2$, the number A_i is also a double root of $\mathcal{P}_{17,i}$. Consequently, for both $i = 1, 2$, the set of numbers $\{\overline{w}_{il},\ l = 1, \ldots, 16\} \cup \{A_i\}$ consists of roots of $\mathcal{P}_{17,i}$ with total multiplicity 18. This is possible only when $\mathcal{P}_{17,i}$, $i = 1, 2$, are trivial polynomials.

In the following computations we make use of the inequalities

$$\mu \neq 0, \qquad \widetilde{\mu} \neq 0, \qquad c_i^2 - a_0 \neq 0, \qquad c_i^2 - \widetilde{a}_0 \neq 0, \qquad c_i^2 - a_1 \neq 0,$$
$$c_i^2 - \widetilde{a}_1 \neq 0, \qquad \alpha \neq 0, \qquad \vartheta \neq 0,$$

following from Lemma 6.3 and (3.43), and the fact that $\nu \neq 0$. Firstly, the coefficients of the term w^{17} in $\mathcal{P}_{17,i}$ yield the equation

$$\frac{3}{2}\delta^{3/2}\{\nu\mu^3\widetilde{\alpha}\widetilde{\mu}^2 - \widetilde{\nu}\widetilde{\mu}^3\alpha\mu^2\} = 0.$$

This implies that

$$\nu\mu\widetilde{\alpha} = \widetilde{\nu}\widetilde{\mu}\alpha. \tag{7.102}$$

The terms w^{16} and w^{15} do not provide additional equations because under the condition (7.102) they vanish automatically. But the coefficients of w^{14} give the equations

$$\frac{9}{2}\delta^{3/2}\nu(c_i^2-a_0)\mu^2\widetilde{\mu}^2\widetilde{\alpha}+2\delta^{3/2}\nu\mu^3\{3\widetilde{\alpha}(c_i^2-\widetilde{a}_0)\widetilde{\mu}+2\widetilde{\vartheta}\widetilde{\mu}\}$$
$$-\frac{9}{2}\delta^{3/2}\widetilde{\nu}(c_i^2-\widetilde{a}_0)\widetilde{\mu}^2\mu^2\alpha-2\delta^{3/2}\widetilde{\nu}\widetilde{\mu}^3\{3\alpha(c_i^2-a_0)\mu+2\vartheta\mu\}=0, \quad i=1,2.$$

Observing (7.102) we deduce from these equations the system

$$\frac{3}{2}c_i^2(\mu-\widetilde{\mu})+\frac{3}{2}(\widetilde{\mu}a_0-\mu\widetilde{a}_0)+4\left(\mu\frac{\widetilde{\vartheta}}{\widetilde{\alpha}}-\widetilde{\mu}\frac{\vartheta}{\alpha}\right)=0, \quad i=1,2.$$

Since $c_1^2\neq c_2^2$ we get from this system

$$\widetilde{\mu}=\mu, \qquad \frac{3}{2}(a_0-\widetilde{a}_0)+4\left(\frac{\widetilde{\vartheta}}{\widetilde{\alpha}}-\frac{\vartheta}{\alpha}\right)=0. \tag{7.103}$$

Further, the coefficients of w^{13} provide the equations

$$\frac{243}{8}\delta^{9/2}\mu^6\widetilde{\mu}^6\left[\nu(c_i^2-\widetilde{a}_1)\mu-\widetilde{\nu}(c_i^2-a_1)\widetilde{\mu}\right]$$
$$\times\left[(c_i^2-a_1)\widetilde{\alpha}-(c_i^2-\widetilde{a}_1)\alpha\right]^2=0, \quad i=1,2. \tag{7.104}$$

By (7.102),

$$\nu(c_i^2-\widetilde{a}_1)\mu-\widetilde{\nu}(c_i^2-a_1)\widetilde{\mu}=\frac{\nu\mu}{\alpha}\left[(c_i^2-\widetilde{a}_1)\alpha-(c_i^2-a_1)\widetilde{\alpha}\right].$$

Thus, from (7.104) we obtain the system

$$c_i^2(\widetilde{\alpha}-\alpha)+\widetilde{a}_1\alpha-a_1\widetilde{\alpha}=0, \quad i=1,2.$$

This implies that

$$\widetilde{\alpha}=\alpha, \qquad \widetilde{a}_1=a_1. \tag{7.105}$$

Now we can resolve (7.102):

$$\widetilde{\nu}=\nu. \tag{7.106}$$

Finally, the coefficients of the free terms in $\mathcal{P}_{17,i}$ yield the equations

$$486\delta^{5/2}(c_i^2-a_0)^4(c_i^2-\widetilde{a}_0)^4\left[\nu(c_i^2-a_0)(c_i^2-\widetilde{a}_1)-\widetilde{\nu}(c_i^2-\widetilde{a}_0)(c_i^2-a_1)\right]$$
$$\times\left[(c_i^2-a_0)(c_i^2-a_1)(\widetilde{\alpha}(c_i^2-\widetilde{a}_0)+\widetilde{\vartheta})\right.$$
$$\left.-(c_i^2-\widetilde{a}_0)(c_i^2-\widetilde{a}_1)(\alpha(c_i^2-a_0)+\vartheta)\right]^2=0, \quad i=1,2.$$

Due to (7.105) and (7.106) this implies that

$$(\widetilde{a}_0 - a_0)\left[c_i^2(\widetilde{\vartheta} - \vartheta) + \widetilde{a}_0\vartheta - a_0\widetilde{\vartheta}\right]^2 = 0, \quad i = 1, 2.$$

Two cases may occur: either $\widetilde{a}_0 - a_0 = 0$ or

$$c_i^2(\widetilde{\vartheta} - \vartheta) + \widetilde{a}_0\vartheta - a_0\widetilde{\vartheta} = 0, \quad i = 1, 2. \tag{7.107}$$

In the case $\widetilde{a}_0 - a_0 = 0$, by means of the second relation in (7.103) and the first relation in (7.105), we have

$$\widetilde{a}_0 = a_0, \qquad \widetilde{\vartheta} = \vartheta. \tag{7.108}$$

In the case (7.107) we also deduce (7.108). Now the proved equalities (7.103), (7.105), (7.106) and (7.108) show that $\widetilde{S} = S$. The proof is complete. □

Proof of Theorem 7.8 Let us suppose that IP6 has two solutions $S = (a_0, a_1, \vartheta, \alpha, \mu)$ and $\widetilde{S} = (\widetilde{a}_0, \widetilde{a}_1, \widetilde{\vartheta}, \widetilde{\alpha}, \widetilde{\mu})$. The uniqueness proof consists of two steps.

Step 1. We prove that the five amplitudes $A_1, \dots, A_5$ uniquely recover the quantities a_0, μ and $\frac{\vartheta}{\alpha}$. Using the formula (7.29) we write amplitude equations for S and $\widetilde{S}$ and extract the velocities therein:

$$c_i^4 - c_i^2\left(2a_0 + \mu A_i - \frac{\vartheta}{\alpha}\right) + \left(a_0 + \frac{\mu}{2}A_i\right)^2 - \left(a_0 + \frac{2}{3}\mu A_i\right)\frac{\vartheta}{\alpha} = 0, \tag{7.109}$$

$$c_i^4 - c_i^2\left(2\widetilde{a}_0 + \widetilde{\mu} A_i - \frac{\widetilde{\vartheta}}{\widetilde{\alpha}}\right) + \left(\widetilde{a}_0 + \frac{\widetilde{\mu}}{2}A_i\right)^2 - \left(\widetilde{a}_0 + \frac{2}{3}\widetilde{\mu} A_i\right)\frac{\widetilde{\vartheta}}{\widetilde{\alpha}} = 0, \tag{7.110}$$

where $i = 1, \dots, 5$. Let us eliminate the velocities. Firstly, we subtract (7.110) from (7.109) and square:

$$\begin{aligned} c_i^4&\left[2(\widetilde{a}_0 - a_0) + (\widetilde{\mu} - \mu)A_i - \frac{\widetilde{\vartheta}}{\widetilde{\alpha}} + \frac{\vartheta}{\alpha}\right]^2 - \left[\left(\widetilde{a}_0 + \frac{\widetilde{\mu}}{2}A_i\right)^2\right. \\ &\left. - \left(a_0 + \frac{\mu}{2}A_i\right)^2 - \left(\widetilde{a}_0 + \frac{2}{3}\widetilde{\mu}A_i\right)\frac{\widetilde{\vartheta}}{\widetilde{\alpha}} + \left(a_0 + \frac{2}{3}\mu A_i\right)\frac{\vartheta}{\alpha}\right]^2 = 0 \end{aligned} \tag{7.111}$$

where $i = 1, \dots, 5$. Secondly, we multiply (7.109) by $2\widetilde{a}_0 + \widetilde{\mu}A_i - \frac{\widetilde{\vartheta}}{\widetilde{\alpha}}$, (7.110) by $2a_0 + \mu A_i - \frac{\vartheta}{\alpha}$ and subtract:

$$\begin{aligned} c_i^4&\left[2(\widetilde{a}_0 - a_0) + (\widetilde{\mu} - \mu)A_i - \frac{\widetilde{\vartheta}}{\widetilde{\alpha}} + \frac{\vartheta}{\alpha}\right] \\ &- \left[\left(\widetilde{a}_0 + \frac{\widetilde{\mu}}{2}A_i\right)^2 - \left(\widetilde{a}_0 + \frac{2}{3}\widetilde{\mu}A_i\right)\frac{\widetilde{\vartheta}}{\widetilde{\alpha}}\right]\left(2a_0 + \mu A_i - \frac{\vartheta}{\alpha}\right) \\ &+ \left[\left(a_0 + \frac{\mu}{2}A_i\right)^2 - \left(a_0 + \frac{2}{3}\mu A_i\right)\frac{\vartheta}{\alpha}\right]\left(2\widetilde{a}_0 + \widetilde{\mu}A_i - \frac{\widetilde{\vartheta}}{\widetilde{\alpha}}\right) = 0 \end{aligned} \tag{7.112}$$

where $i = 1, \dots, 5$. Finally, multiplying (7.112) by $2(\widetilde{a}_0 - a_0) + (\widetilde{\mu} - \mu)A_i - \frac{\widetilde{\vartheta}}{\widetilde{\alpha}} + \frac{\vartheta}{\alpha}$ and subtracting from (7.111), the velocities are eliminated:

$$\Bigg[\Big(\widetilde{a}_0 + \frac{\widetilde{\mu}}{2}A_i\Big)^2 - \Big(a_0 + \frac{\mu}{2}A_i\Big)^2 - \Big(\widetilde{a}_0 + \frac{2}{3}\widetilde{\mu}A_i\Big)\frac{\widetilde{\vartheta}}{\widetilde{\alpha}} + \Big(a_0 + \frac{2}{3}\mu A_i\Big)\frac{\vartheta}{\alpha}\Bigg]^2$$
$$- \Bigg\{\Bigg[\Big(\widetilde{a}_0 + \frac{\widetilde{\mu}}{2}A_i\Big)^2 - \Big(\widetilde{a}_0 + \frac{2}{3}\widetilde{\mu}A_i\Big)\frac{\widetilde{\vartheta}}{\widetilde{\alpha}}\Bigg]\Big(2a_0 + \mu A_i - \frac{\vartheta}{\alpha}\Big)$$
$$- \Bigg[\Big(a_0 + \frac{\mu}{2}A_i\Big)^2 - \Big(a_0 + \frac{2}{3}\mu A_i\Big)\frac{\vartheta}{\alpha}\Bigg]\Big(2\widetilde{a}_0 + \widetilde{\mu}A_i - \frac{\widetilde{\vartheta}}{\widetilde{\alpha}}\Big)\Bigg\}$$
$$\times \Bigg[2(\widetilde{a}_0 - a_0) + (\widetilde{\mu} - \mu)A_i - \frac{\widetilde{\vartheta}}{\widetilde{\alpha}} + \frac{\vartheta}{\alpha}\Bigg], \quad i = 1, \dots, 5.$$

We see that $A_1, \dots, A_5$ are roots of the following polynomial of the 4th degree:

$$\mathcal{P}_4(A) = \Bigg[\Big(\widetilde{a}_0 + \frac{\widetilde{\mu}}{2}A\Big)^2 - \Big(a_0 + \frac{\mu}{2}A\Big)^2 - \Big(\widetilde{a}_0 + \frac{2}{3}\widetilde{\mu}A\Big)\frac{\widetilde{\vartheta}}{\widetilde{\alpha}} + \Big(a_0 + \frac{2}{3}\mu A\Big)\frac{\vartheta}{\alpha}\Bigg]^2$$
$$- \Bigg\{\Bigg[\Big(\widetilde{a}_0 + \frac{\widetilde{\mu}}{2}A\Big)^2 - \Big(\widetilde{a}_0 + \frac{2}{3}\widetilde{\mu}A\Big)\frac{\widetilde{\vartheta}}{\widetilde{\alpha}}\Bigg]\Big(2a_0 + \mu A - \frac{\vartheta}{\alpha}\Big)$$
$$- \Bigg[\Big(a_0 + \frac{\mu}{2}A\Big)^2 - \Big(a_0 + \frac{2}{3}\mu A\Big)\frac{\vartheta}{\alpha}\Bigg]\Big(2\widetilde{a}_0 + \widetilde{\mu}A - \frac{\widetilde{\vartheta}}{\widetilde{\alpha}}\Big)\Bigg\}$$
$$\times \Bigg[2(\widetilde{a}_0 - a_0) + (\widetilde{\mu} - \mu)A - \frac{\widetilde{\vartheta}}{\widetilde{\alpha}} + \frac{\vartheta}{\alpha}\Bigg].$$

Since these roots are different, the polynomial $\mathcal{P}_4$ is trivial.

The coefficient of the term A^4 in $\mathcal{P}_4$ gives the equation $\frac{1}{16}(\widetilde{\mu} - \mu)^4 = 0$. Thus, $\widetilde{\mu} = \mu$. By this equality, the coefficient of A^3 in $\mathcal{P}_4$ vanishes and the coefficient of A^2 yields the equation $\frac{8\mu^2}{9}(\frac{\widetilde{\vartheta}}{\widetilde{\alpha}} - \frac{\vartheta}{\alpha})^2 = 0$. Therefore, $\frac{\widetilde{\vartheta}}{\widetilde{\alpha}} = \frac{\vartheta}{\alpha}$. Finally, by the proved inequalities $\widetilde{\mu} = \mu$ and $\frac{\widetilde{\vartheta}}{\widetilde{\alpha}} = \frac{\vartheta}{\alpha}$ the coefficient of the term A^1 in $\mathcal{P}_4$ provides the equation $\frac{2\mu}{3}\frac{\vartheta}{\alpha}(\widetilde{a}_0 - a_0)^2 = 0$. This implies that $\widetilde{a}_0 = a_0$. Consequently, we have proved the following relations:

$$\widetilde{\mu} = \mu, \qquad \frac{\widetilde{\vartheta}}{\widetilde{\alpha}} = \frac{\vartheta}{\alpha}, \qquad \widetilde{a}_0 = a_0. \tag{7.113}$$

Step 2. We prove that the two points P_j, $j = 1, 2$, contain enough information to reconstruct α and a_1. Since the waves are symmetric when $\nu = 0$, it makes no difference, which side of the extremum P_j, $j = 1, 2$, are located at. Let this be the front side. This means that these points provide the following equations:

$$\xi^+[S, c_j](w_j) = \xi^+[\widetilde{S}, c_j](w_j) = \xi_j, \quad j = 1, 2.$$

Since $\xi^+[S,c_j](A_j)=\xi^+[\widetilde{S},c_j](A_j)=0$, $j=1,2$, by Rolle's theorem there exist $\overline{w}_j\in(w_j,A_j)$, $j=1,2$, such that

$$\xi^+[S,c_j]'(\overline{w}_j)=\xi^+[\widetilde{S},c_j]'(\overline{w}_j)=\xi_j',\quad j=1,2$$

where ξ_j', $j=1,2$, are some numbers. Using these equations in (7.31) and taking the relations $\nu=0$ and (7.113) into account, we deduce the following linear homogeneous system for the quantities $\frac{1}{\widetilde{\alpha}}-\frac{1}{\alpha}$ and $\frac{\widetilde{a}_1}{\widetilde{\alpha}}-\frac{a_1}{\alpha}$:

$$3\delta\left(c_j^2-a_0-\mu\overline{w}_j\right)^2\xi_j'\left[c_j^2\left(\frac{1}{\widetilde{\alpha}}-\frac{1}{\alpha}\right)-\frac{\widetilde{a}_1}{\widetilde{\alpha}}+\frac{a_1}{\alpha}\right]=0,\quad j=1,2.$$

The determinant of this system equals

$$9\delta^2\left(c_2^2-c_1^2\right)\prod_{j=1}^{2}\left(c_j^2-a_0-\mu\overline{w}_j\right)^2\xi_j'.$$

It is different from zero because $c_1^2\neq c_2^2$, $\xi_j'\neq 0$ and $c_j^2-a_0-\mu\overline{w}_j\neq 0$ (cf. Lemma 6.3). This proves that $\frac{1}{\widetilde{\alpha}}=\frac{1}{\alpha}$ and $\frac{\widetilde{a}_1}{\widetilde{\alpha}}=\frac{a_1}{\alpha}$. Combining this result with (7.113) we obtain $\widetilde{S}=S$. The theorem is proved. □

Proof of Theorem 7.9 Let IP7 have two solutions S and $\widetilde{S}$. Again, the proof consists of 2 steps. The first step is identical to the first step in the proof of Theorem 7.8.

Step 2. We are going to show that the given points P_{1l}, $l=1,2$, and P_2 uniquely recover α, a_1 and ν. In the present case we have the following system:

$$\begin{aligned}
&\xi^+[S,c_1](w_{11})=\xi^+[\widetilde{S},c_1](w_{11})=\xi_{11},\\
&\xi^-[S,c_1](w_{12})=\xi^-[\widetilde{S},c_1](w_{12})=\xi_{12},\\
&\xi[S,c_2](w_2)=\xi[\widetilde{S},c_2](w_2)=\xi_2
\end{aligned}$$

where $\xi[S,c_2]$ is either $\xi^+[S,c_2]$ or $\xi^-[S,c_2]$, depending on the location of P_2. Using the corresponding amplitude relations and Rolle's theorem, as before, we conclude that there exist $\overline{w}_{11}\in(w_{11},A_1)$, $\overline{w}_{12}\in(w_{12},A_1)$ and $\overline{w}_2\in(w_2,A_2)$ such that

$$\begin{aligned}
&\xi^+[S,c_1]'(\overline{w}_{11})=\xi^+[\widetilde{S},c_1]'(\overline{w}_{11})=\xi_{11}',\\
&\xi^-[S,c_1]'(\overline{w}_{12})=\xi^-[\widetilde{S},c_1]'(\overline{w}_{12})=\xi_{12}',\\
&\xi[S,c_2]'(\overline{w}_2)=\xi[\widetilde{S},c_2]'(\overline{w}_2)=\xi_2'
\end{aligned}$$

with some numbers ξ_{1l}', $l=1,2$ and ξ_2'. Applying these relations in (7.31) and observing (7.113), we reach the following linear homogeneous system for the quantities $\frac{1}{\widetilde{\alpha}}-\frac{1}{\alpha}$, $\frac{\widetilde{a}_1}{\widetilde{\alpha}}-\frac{a_1}{\alpha}$ and $\frac{\widetilde{\nu}}{\widetilde{\alpha}}-\frac{\nu}{\alpha}$:

$$3\delta\left(c_1^2 - a_0 - \mu\overline{w}_{1l}\right)^2 \xi'_{1l}\left[c_1^2\left(\frac{1}{\widetilde{\alpha}} - \frac{1}{\alpha}\right) - \frac{\widetilde{a}_1}{\widetilde{\alpha}} + \frac{a_1}{\alpha}\right]$$

$$- 2\delta^{3/2}\left(c_1^2 - a_0 - \mu\overline{w}_{1l}\right)^3\left(\frac{\widetilde{\nu}}{\widetilde{\alpha}} - \frac{\nu}{\alpha}\right) = 0, \quad l = 1, 2,$$

$$3\delta\left(c_2^2 - a_0 - \mu\overline{w}_2\right)^2 \xi'_2\left[c_2^2\left(\frac{1}{\widetilde{\alpha}} - \frac{1}{\alpha}\right) - \frac{\widetilde{a}_1}{\widetilde{\alpha}} + \frac{a_1}{\alpha}\right]$$

$$- 2\delta^{3/2}\left(c_2^2 - a_0 - \mu\overline{w}_2\right)^3\left(\frac{\widetilde{\nu}}{\widetilde{\alpha}} - \frac{\nu}{\alpha}\right) = 0.$$

The determinant is

$$18\delta^{7/2}\xi'_2\left(c_1^2 - c_2^2\right)\left(c_2^2 - a_0 - \mu\overline{w}_2\right)\prod_{l=1}^{2}\left(c_1^2 - a_0 - \mu\overline{w}_{1l}\right)$$

$$\times\left\{\left(c_1^2 - a_0 - \mu\overline{w}_{11}\right)\xi'_{12} - \left(c_1^2 - a_0 - \mu\overline{w}_{12}\right)\xi'_{11}\right\}.$$

Again, the determinant is different from zero because $c_1^2 \neq c_2^2$, $\xi'_{1l}, \xi'_2 \neq 0$, the factors ξ'_{11} and ξ'_{12} have different signs (since P_{11} and P_{12} stand on different sides of the extremum), the quantities of the type $c^2 - a_0 - \mu w$ are different from zero (by Lemma 6.3) and $c_1^2 - a_0 - \mu\overline{w}_{11}$, $c_1^2 - a_0 - \mu\overline{w}_{12}$ have a common sign. The latter statement simply follows from the inequality $c_1^2 - a_0 - \mu w \neq 0$, $w \in (0, A_1]$, that is satisfied for the first solitary wave solution by Lemma 6.3. Therefore, the solution of the regular homogeneous system under consideration is $\frac{1}{\widetilde{\alpha}} - \frac{1}{\alpha} = \frac{\widetilde{a}_1}{\widetilde{\alpha}} - \frac{a_1}{\alpha} = \frac{\widetilde{\nu}}{\widetilde{\alpha}} - \frac{\nu}{\alpha} = 0$. These relations with (7.113) imply that $\widetilde{S} = S$. The proof is complete. □

Proofs of Theorems 7.10 and 7.11 Let IP8 (resp. IP9) have two solutions $S = (a_1, \alpha, \vartheta)$ and $\widetilde{S} = (\widetilde{a}_1, \widetilde{\alpha}, \widetilde{\vartheta})$ (resp. $S = (a_1, \alpha, \vartheta, \nu)$ and $\widetilde{S} = (\widetilde{a}_1, \widetilde{\alpha}, \widetilde{\vartheta}, \widetilde{\nu})$). In the first step of the proofs of these theorems we make use of the formula (7.29) to get the single-valued explicit expression (7.44) for $\frac{\widetilde{\vartheta}}{\widetilde{\alpha}} = \frac{\vartheta}{\alpha}$. The second steps of the proofs of Theorems 7.10 and 7.11 repeat the second steps of the proofs of Theorems 7.8 and 7.9, respectively. □

Chapter 8
Summary

8.1 General Glance at Mathematical Methods

In this chapter we discuss the mathematical methods developed in the book from a more general viewpoint.

Let us start with the generalisation of the material of Chap. 5. Let us be given a linear dispersive equation of motion or system of equations of motion containing coefficients $S=(s_1,\ldots,s_m)$ to be determined. Suppose that the dispersion relation of this model is of the polynomial form

$$\mathcal{P}(\omega,k)=0 \quad \text{where } \mathcal{P}(\omega,k)=\sum_{i_1=0}^{n_1}\sum_{i_2=0}^{n_2}\varkappa_{i_1 i_2}\omega^{i_1}k^{i_2}.$$

Then the reconstruction of S is split into two steps.

(1) Determination of the vector of coefficients $K=(\varkappa_{i_1 i_2})_{\substack{i_1=0,\ldots,n_1\\ i_2=0,\ldots,n_2}}$ of the polynomial $\mathcal{P}$ from frequency-wavenumber pairs (ω_j,k_j), $k=1,\ldots,n$. This means the solution of the linear system of equations

$$\sum_{i_1=0}^{n_1}\sum_{i_2=0}^{n_2}\omega_j^{i_1}k_j^{i_2}\cdot\varkappa_{i_1 i_2}=0, \quad j=1,\ldots,n.$$

(2) Computation of $s_1,\ldots,s_m$ by means of obtained coefficients $\varkappa_{i_1 i_2}$. Usually this consists in solving a nonlinear system of functional equations

$$\mathcal{R}(K,S)=0$$

where $\mathcal{R}$ represents the relations between the coefficients of the original problem and the dispersion equation.

The frequency-wavenumber pairs (ω_j,k_j) can be obtained in a different manner. If possible, one can directly measure harmonic waves. But a more general way is the

J. Janno, J. Engelbrecht, *Microstructured Materials: Inverse Problems*, Springer Monographs in Mathematics,
DOI 10.1007/978-3-642-21584-1_8,

spectral decomposition of a linear wave packet and extraction of the pairs (ω_j, k_j) from the spectrum. The latter method was described in detail in Sect. 5.3.

Another possibility is to use phase and group velocities $c_{ph,j}, c_{g,j}$ and the dispersion parameters d_j of Gaussian wave packets with central frequencies ω_j. Such packets are especially easy to generate and measure in practice. The given data again provide information about the derivatives of the dispersion function: $k_j = k(\omega_j) = \frac{\omega_j}{c_{ph,j}}$, $k'_j = k'(\omega_j) = \frac{1}{c_{g,j}}$, $k''(\omega_j) = 2d_j$ (cf. Sect. 4.2.3). When the phase and group velocities $c_{ph,j}$ and $c_{g,j}$ of l packets are measured then the first step of reconstruction is as follows.

(1) Solution of the following linear system of equations that is derived from formulas for $\mathcal{P}$ and $\frac{\partial}{\partial\omega}\mathcal{P}$:

$$\sum_{i_1=0}^{n_1}\sum_{i_2=0}^{n_2} \omega_j^{i_1} k_j^{i_2} \cdot \varkappa_{i_1 i_2} = 0, \quad j = 1, \dots, l,$$

$$\sum_{i_1=0}^{n_1}\sum_{i_2=0}^{n_2} \left[i_1 \omega_j^{i_1-1} k_j^{i_2} + i_2 \omega_j^{i_1} k_j^{i_2-1} k'_j\right] \cdot \varkappa_{i_1 i_2} = 0, \quad j = 1, \dots, l$$

for the vector $K = (\varkappa_{i_1 i_2})_{\substack{i_1=0,\dots,n_1 \\ i_2=0,\dots,n_2}}$.

But when the phase and group velocities $c_{ph,j}$ and $c_{g,j}$ and the dispersion parameters d_j of q packets are measured then the first step of reconstruction is as follows.

(1) Solution of the following linear system of equations that is derived from formulas for $\mathcal{P}$, $\frac{\partial}{\partial\omega}\mathcal{P}$ and $\frac{\partial^2}{\partial\omega^2}\mathcal{P}$:

$$\sum_{i_1=0}^{n_1}\sum_{i_2=0}^{n_2} \omega_j^{i_1} k_j^{i_2} \cdot \varkappa_{i_1 i_2} = 0, \quad j = 1, \dots, q,$$

$$\sum_{i_1=0}^{n_1}\sum_{i_2=0}^{n_2} \left[i_1 \omega_j^{i_1-1} k_j^{i_2} + i_2 \omega_j^{i_1} k_j^{i_2-1} k'_j\right] \cdot \varkappa_{i_1 i_2} = 0, \quad j = 1, \dots, q,$$

$$\sum_{i_1=0}^{n_1}\sum_{i_2=0}^{n_2} \Big[i_1(i_1-1)\omega_j^{i_1-2} k_j^{i_2} + 2 i_1 i_2 \omega_j^{i_1-1} k_j^{i_2-1} k'_j$$
$$+ i_1(i_2-1)\omega_j^{i_1-1} k_j^{i_2-2} (k'_j)^2 + i_2 \omega_j^{i_1} k_j^{i_2-1} k''_j\Big] \cdot \varkappa_{i_1 i_2} = 0, \quad j = 1, \dots, q$$

for the vector $K = (\varkappa_{i_1 i_2})_{\substack{i_1=0,\dots,n_1 \\ i_2=0,\dots,n_2}}$.

The second step remains the same as before.

To prove regularity of the linear systems in the first step, the method of vanishing polynomial coefficients can be used.

Now we proceed to the generalisation of some material of Chap. 7. If the dispersive wave equation or system is nonlinear, then in the case of a proper balance between the dispersion and nonlinearity the solitary waves emerge. Those waves can be measured and used to reconstruct the coefficients $S=(s_1,\ldots,s_m)$ of the wave equation or system under consideration. The mean value theorems can be used to study the uniqueness and stability of the solutions of the inverse problems for solitary waves. Let us explain how this can be done in the general case. To use these theorems, an ordinary differential equation that describes the solitary wave process must be autonomous, have been integrated up to the order 1 and have a polynomial form. Such an equation can be written in the form

$$\mathcal{Q}(\xi', w) = 0 \tag{8.1}$$

where $\xi(w)$ is an inverse of the solitary wave function $w(\xi)$ and $\mathcal{Q}$ is a polynomial. To emphasise the dependence on S, let us write $\xi[S](w)$.

Suppose that a solitary wave is measured at some point $P(\xi_1, w_1)$ and the amplitude A is also known (i.e. $w(0) = A$). Then one has the nonlinear equation $\xi[S](w_1) = \xi_1$. When many measurements of solitary waves are provided, a system of such equations is formed to determine S. In general, such a system is quite complicated and may contain higher transcendental functions. The Rolle's mean value theorem enables us to reduce the uniqueness of the solution of this system to the uniqueness of the solution of an algebraic system. Clearly, the latter is easier to handle.

Assume that $\widetilde{S}$ is also a solution of the inverse problem under consideration. Then

$$\xi[S](w_1) = \xi_1 = \xi[\widetilde{S}](w_1).$$

Since $\xi[S](A) = \xi[\widetilde{S}](A) = 0$, by Rolle's theorem, there exists a number $\overline{w}_1$ between A and w_1 such that

$$\xi[S]'(\overline{w}_1) = \xi[\widetilde{S}]'(\overline{w}_1) =: \xi_1'.$$

Insertion of the numbers $\overline{w}_1$ and ξ_1' into (8.1) yields

$$\mathcal{Q}(\xi_1', \overline{w}_1) = 0.$$

This equation is algebraic with respect to S, because $\mathcal{Q}$ is a polynomial and the coefficients of $\mathcal{Q}$ are algebraic expressions of the components of S. Since $\widetilde{S}$ is also the solution of the inverse problem, the same algebraic equation holds with two different sets of coefficients. In case many measurements of solitary waves are provided, a whole system of such equations is formed. Thus, the uniqueness for the original inverse problem is reduced to the uniqueness for an algebraic coefficient-type problem.

8.2 From Mathematics to Physics

Chapters 5 and 7 dealt with inverse problems from the mathematical viewpoint. The problems were set up either for the hierarchical equation or for the coupled system in the dimensionless form. Upon the solution of these problems, it is possible to reconstruct the original physical constants, too. Here we explain how this can be done.

For better readability we repeat here the basic equations and relations from Sects. 3.2 and 3.3. The potential energy function for a microstructured material in the linear setting is

$$W = \frac{1}{2}aU_X^2 + A\varphi U_X + \frac{1}{2}B\varphi^2 + \frac{1}{2}C\varphi_X^2$$

and in the nonlinear case reads

$$W = \frac{1}{2}aU_X^2 + A\varphi U_X + \frac{1}{2}B\varphi^2 + \frac{1}{2}C\varphi_X^2 + \frac{1}{6}NU_X^3 + \frac{1}{6}M\varphi_X^3.$$

The final dimensionless mathematical models are the following.

- The linear hierarchical equation

$$v_{tt} = bv_{xx} + \delta(\beta v_{tt} - \gamma v_{xx})_{xx}.$$

- The nonlinear hierarchical equation

$$v_{tt} = bv_{xx} + \frac{\mu}{2}\left(v^2\right)_{xx} + \delta(\beta v_{tt} - \gamma v_{xx})_{xx} + \delta^{3/2}\frac{\lambda}{2}\left(v_x^2\right)_{xxx}.$$

- The linear coupled system

$$v_{tt} = a_0 v_{xx} + \vartheta_0 \varphi_{xx},$$
$$\delta\varphi_{tt} = \delta a_1 \varphi_{xx} - \alpha\varphi - \vartheta_1 v.$$

- The nonlinear coupled system

$$v_{tt} = a_0 v_{xx} + \frac{\mu}{2}\left(v^2\right)_{xx} + \vartheta_0 \varphi_{xx},$$
$$\delta\varphi_{tt} = \delta a_1 \varphi_{xx} + \delta^{3/2}\nu_1 \varphi_x \varphi_{xx} - \alpha\varphi - \vartheta_1 v.$$

The physical parameters included in the energy functions are a, A, B, C (linear case) and a, A, B, C, N, M (nonlinear case). In addition, from the Euler–Lagrange equations (3.7) and (3.8) we have the constants ρ_0 and I, and the dimensionless models contain the geometrical parameters

$$\delta = \frac{l^2}{L^2}, \qquad \varepsilon = \frac{U_0}{L}, \qquad \varkappa = \frac{T_0^2}{L^2}$$

where l is the scale of the microstructure and U_0, L and T_0 are certain constants (e.g. the fixed amplitude, wavelength and period, respectively).

First we note that

$$a_0 = \frac{a\varkappa}{\rho_0}, \qquad a_1 = \frac{C^*}{I^*} \tag{8.2}$$

where C^* and I^* are related to C and I by

$$C^* = \frac{C}{l^2}, \qquad I^* = \frac{I}{\varkappa l^2}. \tag{8.3}$$

For the linear hierarchical equation we have

$$b = \frac{a\varkappa}{\rho_0}\left(1 - \frac{A^2}{aB}\right), \qquad \beta = \frac{A^2 \varkappa I^*}{B^2 \rho_0}, \qquad \gamma = \frac{A^2 \varkappa C^*}{B^2 \rho_0} \tag{8.4}$$

and for the nonlinear hierarchical equation in addition

$$\mu = \frac{N\varkappa\varepsilon}{\rho_0}, \qquad \lambda = \frac{A^3 M^* \varkappa\varepsilon}{B^3 \rho_0}. \tag{8.5}$$

Further, in the coupled system the following additional parameters occur in the linear case

$$\vartheta_0 = \frac{A\varkappa}{\varepsilon\rho_0}, \qquad \alpha = \frac{B}{I^*}, \qquad \vartheta_1 = \frac{A\varepsilon}{I^*} \tag{8.6}$$

and in the nonlinear case

$$\nu_1 = \frac{M^*}{I^*} \quad \text{with } M^* = \frac{M}{l^3}. \tag{8.7}$$

The mathematical inverse problems provide

$$\vartheta = \vartheta_0 \vartheta_1 = \frac{A^2 \varkappa}{\rho_0 I^*}, \qquad \nu = \frac{\nu_1}{\vartheta_0} = \frac{M^* \varepsilon\rho_0}{I^* A\varkappa} \tag{8.8}$$

instead of ϑ_0, ϑ_1 and ν_1.

As mentioned, the geometric parameters l, δ, ε, $\varkappa$ are assumed to be known. The original physical parameters can be partially reconstructed as follows:

(i) Linear case, hierarchical equation. The solution of the mathematical inverse problem is the triplet b, β, γ. Assuming that ρ_0, I and a are additionally known, it is possible to determine the physical parameters B and C and the absolute value of A. They can be computed by the following formulas that are deduced from (8.3) and (8.4):

$$B = \frac{I(\varkappa a - \rho_0 b)}{\varkappa l^2 \rho_0 \beta}, \qquad |A| = \sqrt{\left(a - \frac{\rho_0 b}{\varkappa}\right) B}, \qquad C = \frac{l^2 \rho_0 \gamma B^2}{\varkappa A^2}. \tag{8.9}$$

(ii) Linear case, coupled system. The inverse problem yields the 4-vector with components $a_0, a_1, \alpha, \vartheta$. Suppose that ρ_0 and I are known. Then from (8.2), (8.3) and (8.6) the parameters a, B, C and $|A|$ can be obtained. The formulas are as follows.

$$a = \frac{\rho_0 a_0}{\varkappa}, \qquad |A| = \sqrt{\frac{\rho_0 I \vartheta}{\varkappa^2 l^2}}, \qquad B = \frac{I\alpha}{\varkappa l^2}, \qquad C = \frac{I a_1}{\varkappa}. \tag{8.10}$$

(iii) Nonlinear case, hierarchical equation. The solution of the inverse problem is the 5-vector with components b, μ, β, γ, λ. Assuming again that ρ_0, I and a are known, it is possible to reconstruct the physical parameters B, C, N and the absolute values of A and M. Namely, B, C and $|A|$ are given by (8.9) and formulas for N, $|M|$ follow from (8.5) with (8.7):

$$N = \frac{\rho_0 \mu}{\varkappa \varepsilon}, \qquad |M| = \frac{l^3 \rho_0 |\lambda| B^3}{\varkappa \varepsilon |A|^3}. \tag{8.11}$$

Moreover, it is possible to determine the sign of the product AM. Namely, from the right relation in (8.5) and the physical inequalities (3.15) and $l, \varkappa, \varepsilon > 0$ it follows that

$$\operatorname{sign} AM = \operatorname{sign} \lambda. \tag{8.12}$$

(iv) Nonlinear case, coupled system. The inverse problem IP7 gives the 6-vector with components a_0, a_1, α, ϑ, μ, ν. Again, assume that ρ_0 and I are known. Then it is possible to determine a, B, C, N, $|A|$ and $|M|$. The parameters a, B, C, $|A|$ and N are given by (8.10) and the left-hand relation in (8.11). The formula for $|M|$ can be deduced from (8.8) with (8.3) and (8.7):

$$|M| = \frac{lI|\nu||A|}{\varepsilon \rho_0}. \tag{8.13}$$

In addition, we have the sign relation

$$\operatorname{sign} AM = \operatorname{sign} \nu. \tag{8.14}$$

Let us make some remarks.

(1) The determination of the signs of the parameters A and M from macro-waves is not possible. All related expressions contain these parameters only in the forms A^2, AM or M/A. However, these signs could be reconstructed if additionally the measurement of micro-waves is possible. For instance, the macro- and micro-amplitudes of harmonic waves or solitary waves provide ϑ_0 (see (5.17) and (7.30)). Then from the left-hand relation in (8.6) we easily get A with the right sign and from (8.14) the sign of M, too.
(2) The waves considered in this book do not provide enough information to recover the whole vector of physical parameters ρ_0, I, a, A, B, C, N, M. This is so because we are limited to free waves whose motion is governed by homogeneous equations. Evidently, it is possible to multiply all coefficients of the homogeneous equations (3.18) and (3.19) by a common constant without changing the wave functions U and φ. This means that all vectors of the form $c\rho_0$, cI, ca, cA, cB, cC, cN, cM with arbitrary $c \in \mathbb{R}_+$ fit to any wave measurements. The reconstruction of the whole vector may be expected only in the presence of mass forces when the governing equations are nonhomogeneous.
(3) Instead of the triplet ρ_0, I, a or pair ρ_0, I, other subtriplets or -pairs of the vector ρ_0, I, a, A, B, C, N, M may be chosen as given quantities in cases

(i)–(iv). Then the corresponding subvectors of unknowns also change. Related solution formulas can be deduced in such situations, as well.

It is important to emphasise that the hierarchical equation is not only a simplified version of the coupled system—this model can be deduced by means of different arguments. The suitability of models (e.g. hierarchical equation versus coupled system) for particular materials can be established by means of the solution of inverse problems (for details see Sect. 2.1).

The scale of nonlinearity also depends on the material. Registration of higher harmonics [52] indicates the presence of nonlinear effects. If this is essential, the nonlinear theory should be applied.

8.3 Epilogue

The ideology described in this book is actually a consistent presentation of the theoretical ultrasonic NDE for microstructured materials. It starts from a well-grounded mathematical model; after that the informative physical effects are analysed which are followed by the analysis of the corresponding inverse problems. The important uniqueness and stability issues related to the inverse problems are studied.

The numerical tests show that the sensitivity of different material parameters on the errors of the data varies very much. Such a phenomenon may have the physical explanation: some parameters influence more the micro-process than the macro-process. Reconstruction of sensitive parameters by means of macrodeformation requires precise measurements.

The informative effects which can be found in propagating waves in microstructured solids are: (i) dispersive effects; (ii) nonlinear effects at macro- and micro-level combined with dispersion. The case (i) means that the dependence of phase and/or group velocities on dispersion parameters which reflect the properties of a microstructure, is used. The case (ii) means that the balance of nonlinear and dispersive effects leads to asymmetric solitary waves, and this asymmetry reflects the properties of a microstructure. We have published our findings in a series of research papers [17, 20, 34–39] and here these results are systematically collected and revised. This presentation can serve as an example of a rigorous theoretical analysis which is actually needed for all models used for the NDE.

The mathematical model we have used is based on the Mindlin [50] model elaborated later by Engelbrecht et al. [16, 18]. It has been shown that this model is rather general and can be derived also by using the concept of pseudomomentum [48] and the concept of internal variables [3]. With the full confidence in this model, we stress that it describes a dispersive material and it is quite natural to include physical nonlinearities into the model. Clearly a possible way to enhance the model is to include also viscous effects. We have done it earlier [13, 15] but within a different framework, and the inverse problems based on Mindlin-type viscous models need to be studied more thoroughly. Actually in a general framework, many examples on the decay of ultrasound are given in [43] and [44]. Decay of ultrasound is extremely important in biological tissues [65]. In some cases dissipative effects are most informative as in the case of diagnosis of bacterially infected wood [57]. Frequency-dependent dissipation may be important in rock mechanics [40] and in ultrasonic medical imaging [6]. So there is a wide area for further studies.

Another important aspect also needs clarifying—namely dimensionality. Our theory is one-dimensional but the ultrasound transducers actually generate wave-beams (cf. Chap. 2). Therefore a question about the influence of diffraction effects in the perpendicular direction to the beam axis is clearly justified. These effects are usually described by two-dimensional evolution equations [11] which like the celebrated KdV equation are one-wave equations. Here our model like the classical wave equation is a two-wave equation (or a system). For this model, we have constructed a full theory with needed mathematical proofs. The next step will be to develop a similar approach for evolution equations, especially for a two-dimensional case. For a one-dimensional case, the corresponding evolution equation is already derived by Randrüüt and Braun [55] which explicitly demonstrates the role of nonlinearities on macro- and micro-levels resulting in an asymmetric profile.

Last but not least, the theory waits for applications.

References

1. Anger, G.: Inverse Problems in Differential Equations. Plenum, London (1990)
2. Avriel, M.: Nonlinear Programming. Analysis and Methods. Dover, New York (2003)
3. Berezovski, A., Engelbrecht, J., Maugin, G.A.: Generalized thermomechanics with dual internal variables. Arch. Appl. Mech. **81**, 229–240 (2011)
4. Capriz, G.: Continua with Microstructure. Springer, New York (1989)
5. Cantrell, J.H.: Fundamentals and applications of nonlinear ultrasonic nondestructive evaluation. In: Kundu, T. (ed.) Ultrasonic Nondestructive Evaluation: Engineering and Biological Material Characterization, pp. 363–434. CRC Press, Boca Raton (2004)
6. Chen, W., Zhang, X., Cai, X.: A study on modified Szabo's wave equation modelling of frequency-dependent dissipation in ultrasonic medical imaging. Phys. Scr. T **136**, 014014 (2009)
7. Colton, D., Kress, R.: Inverse Acoustic and Electromagnetic Scattering Theory. Appl. Math. Sci., vol. 93. Springer, New York (1992)
8. Dauxois, T., Peyrard, M.: Physics of Solitons. Cambridge University Press, Cambridge (2006)
9. Delsanto, P.-P.: Universality of Nonclassical Nonlinearity: Applications to Non-destructive Evaluations and Ultrasonics. Springer, New York (2007)
10. Elmore, W.C., Heald, M.A.: Physics of Waves. Dover, New York (1969)
11. Engelbrecht, J.: Nonlinear Wave Processes of Deformation in Solids. Pitman, London (1983)
12. Engelbrecht, J.: Nonlinear Waves Dynamics. Complexity and Simplicity. Kluwer Academic, Dordrecht (1997)
13. Engelbrecht, J., Ravasoo, A.: From continuum mechanics to applications in the nondestructive testing. Bull. Tech. Univ. Istanb. **47**(1–2), 83–103 (1994) (Suhubi Special Issue)
14. Engelbrecht, J., Pastrone, F.: Waves in microstructured solids with strong nonlinearities in microscale. Proc. Est. Acad. Sci., Phys. Math. **52**, 12–20 (2003)
15. Engelbrecht, J., Sillat, T.: Wave propagation in dissipative microstructured materials. Proc. Est. Acad. Sci., Phys. Math. **52**, 103–114 (2003)
16. Engelbrecht, J., Berezovski, A., Pastrone, F., Braun, M.: Waves in microstructured materials and dispersion. Philos. Mag. **85**(33–35), 4127–4141 (2005)
17. Engelbrecht, J., Janno, J.: Microstructured solids and inverse problems. Rend. Semin. Mat. (Torino) **65**, 159–169 (2007)
18. Engelbrecht, J., Pastrone, F., Braun, M., Berezovski, A.: Hierarchies of waves in nonclassical materials. In: Delsanto, P.-P. (ed.) Universality of Nonclassical Nonlinearity: Application to Non-destructive Evaluation and Ultrasonics, pp. 29–47. Springer, New York (2007)
19. Engelbrecht, J., Berezovski, A., Soomere, T.: Highlights in the research into complexity of nonlinear waves. Proc Est. Acad. Sci. **59**, 61–65 (2010)
20. Engelbrecht, J., Ravasoo, A., Janno, J.: Nonlinear acoustic nondestructive evaluation (NDE): qualitative and quantitative effects. Mater. Manuf. Process. **25**, 212–220 (2010)

J. Janno, J. Engelbrecht, *Microstructured Materials: Inverse Problems*,
Springer Monographs in Mathematics,
DOI 10.1007/978-3-642-21584-1,

21. Engl, H.W., Hanke, M., Neubauer, A.: Regularization of Inverse Problems. Kluwer Academic, Dordrecht (1996)
22. Erdélyi, A. (ed.): Tables of Integral Transforms, vol. 1. McGraw-Hill, New York (1954)
23. Eringen, A.C.: Nonlinear Theory of Continuous Media. McGraw-Hill, London (1962)
24. Eringen, A.C.: Linear theory of micropolar elasticity. J. Math. Mech. **15**, 909–923 (1966)
25. Eringen, A.C.: Microcontinuum Field Theories. Foundations and Solids. Springer, New York (1999)
26. Eringen, A.C., Maugin, G.A.: Electrodynamics of Continua, I, II. Springer, New York (1989)
27. Eringen, A.C., Suhubi, E.S.: Elastodynamics I. Academic Press, London (1974)
28. Gates, T.S., Odegard, G.M., Frankland, S.J.V., Clancy, T.C.: Computational materials: multi-scale modeling and simulation of nanostructured materials. Compos. Sci. Technol. **65**, 2416–2434 (2005)
29. Gladwell, G.M.L.: Inverse Problems in Vibration, 2nd edn. Kluwer Academic, Dordrecht (2004)
30. Hadamard, J.: Lectures on Cauchy's Problem in Linear Partial Differential Equations. Dover, New York (1953)
31. Hauk, V.: Structural and Residual Stress Analysis by Nondestructive Methods. Elsevier, Amsterdam (1997)
32. Hellier, C.J.: Handbook of Nondestructive Evaluation. McGraw-Hill, New York (2001)
33. Isakov, V.: Inverse Problems for Partial Differential Equations. Springer, New York (1998)
34. Janno, J., Engelbrecht, J.: Waves in microstructured solids: inverse problems. Wave Motion **43**, 1–11 (2005)
35. Janno, J., Engelbrecht, J.: Solitary waves in nonlinear microstructured materials. J. Phys. A, Math. Gen. **38**, 5159–5172 (2005)
36. Janno, J., Engelbrecht, J.: An inverse solitary wave problem related to microstructured materials. Inverse Probl. **21**, 2019–2034 (2005)
37. Janno, J., Engelbrecht, J.: Determining properties of nonlinear microstructured materials by means of solitary waves. In: Lesnic, D. (ed.) Proc. 5th International Conference on Inverse Problems in Engineering, Cambridge, 11–15 July 2005, vol. II, J02. Leeds Univ. Press, Leeds (2005)
38. Janno, J., Engelbrecht, J.: Inverse problems related to a coupled system of microstructure. Inverse Probl. **24**, 045017 (2008)
39. Janno, J., Engelbrecht, J.: Identification of microstructured materials by phase and group velocities. Math. Model. Anal. **14**, 57–68 (2009)
40. Johnson, P.: Nonequilibrium nonlinear dynamics in solids: state of the art. In: Delsanto, P.-P. (ed.) Universality of Nonclassical Nonlinearity: Application to Non-destructive Evaluations and Ultrasonics, pp. 49–69. Springer, New York (2007)
41. Kabanikhin, S.I., Lorenzi, A.: Identification Problems of Wave Phenomena. Theory and Numerics. VSP, Utrecht (1999)
42. Korteweg, D.J., de Vries, G.: On the change of form of long waves advancing in a rectangular channel and on a new type of long stationary waves. Philos. Mag., Ser 5 **39**, 422–443 (1895)
43. Krautkrämer, J., Krautkrämer, H.: Ultrasonic Testing of Materials., 4th edn. Springer, Berlin (1990)
44. Kundu, T. (ed.): Ultrasonic Nondestructive Evaluation: Engineering and Biological Material Characterization. CRC Press, Boca Raton (2004)
45. Leto, J.A., Choudhury, S.R.: Solitary wave families of a generalized microstructure PDE. Commun. Nonlinear Sci. Numer. Simul. **14**, 1999–2005 (2009)
46. Liu, G.R., Han, X.: Computational Inverse Techniques in Nondestructive Evaluation. CRC Press, London (2003)
47. Luenberger, D.G., Ye, Y.: Linear and Nonlinear Programming, 3rd edn. Springer, New York (2008)
48. Maugin, G.A.: Material Inhomogeneities in Elasticity. Chapman & Hall, London (1993)
49. McGonnagie, W.J.: Nondestructive Testing. McGraw-Hill, New York (1961)
50. Mindlin, R.D.: Micro-structure in linear elasticity. Arch. Ration. Mech. Anal. **16**, 51–78 (1964)

51. Mitchell, M.: An Introduction to Genetic Algorithms. MIT Press, Cambridge (1998)
52. Naugolnykh, K., Ostrovsky, L.: Nonlinear Wave Processes in Acoustics. Cambridge University Press, Cambridge (1998)
53. Pastrone, F.: Nonlinearity and complexity in elastic wave motion. In: Delsanto, P.-P. (ed.) Universality of Nonclassical Nonlinearity: Application to Non-destructive Evaluation and Ultrasonics, pp. 15–26. Springer, New York (2007)
54. Porubov, A.V.: Amplification of Nonlinear Strain Waves in Solids. World Scientific, Singapore (2003)
55. Randrüüt, M., Braun, M.: On one-dimensional solitary waves in microstructured solids. Wave Motion **47**, 217–230 (2010)
56. Romanov, V.G.: Inverse Problems of Mathematical Physics. VNU Science Press, Utrecht (1987)
57. Ross, R.J., Ward, J.C., TenWolde, A.: Identifying bacterially infected oak by stress wave non-destructive evaluation. US DoA, Research Paper FPL-RP-512 (1992)
58. Saks, S., Zygmund, A.: Analytic Functions. Pol. Tow. Mat., Warsaw (1952)
59. Salupere, A., Tamm, K., Engelbrecht, J.: Numerical simulation of interaction of solitary deformation waves in microstructured solids. Int. J. Non-Linear Mech. **43**, 201–208 (2008)
60. Santamarina, J.C., Fratta, D.: Discrete Signals and Inverse Problems. An Introduction for Engineers and Scientists. Wiley, New York (2005)
61. Schneider, E.: Ultrasonic techniques. In: Hauk, V. (ed.) Structural and Residual Stress Analysis by Nondestructive Methods, pp. 522–563. Elsevier, Amsterdam (1997)
62. Scott Russell, J.: Report on waves. In: Fourteenth Meeting of the British Association for the Advancement of Science, pp. 311–390. John Murray, London (1844)
63. Sertakov, I.: Inverse problems in Mindlin's model of microstructure. MSc Thesis. Tallinn UT, Tallinn (2010)
64. Shull, P.J.: Nondestructive Evaluation: Theory, Techniques and Applications. Marcel Dekker, New York (2002)
65. Shung, K.K., Thieme, G.A. (eds.): Ultrasonic Scattering in Biological Tissues. Springer, New York (1992)
66. Taniuti, T., Nishihara, K.: Nonlinear Waves. Pitman, London (1983). In Japanese (1977)
67. Thompson, D.O., Chimenti, D.E. (eds.): Review of Progress in Quantitative Evaluation. Plenum Press, New York (1986)
68. Tikhonov, A.N., Arsenin, V.Ya.: Solution of Ill-Posed Problems. Wiley, New York (1977). Transl. from Russian
69. Truell, R., Elbaum, C., Chick, B.B.: Ultrasonic Methods in Solid State Physics. Academic Press, New York (1969)
70. Trujillo, D.M., Busby, H.R.: Practical Inverse Analysis in Engineering. CRC Press, London (1997)
71. Wells, P.N.T. (ed.): Ultrasonics in Clinical Diagnostics. Churchill Livingstone, Edinburgh (1977)
72. Zabusky, N.J., Kruskal, M.D.: Interaction of "solitons" in a collisionless plasma and the recurrence of initial states. Phys. Rev. Lett. **15**, 240–243 (1965)
73. Zhang, R., Jiang, B., Cao, W.: Influence of sample size on ultrasonic phase velocity measurements in piezoelectric ceramics. J. Appl. Phys. **91**, 10194–10198 (2002)

Index

J. Janno, J. Engelbrecht, *Microstructured Materials: Inverse Problems*,
Springer Monographs in Mathematics,
DOI 10.1007/978-3-642-21584-1, © Springer-Verlag Berlin Heidelberg 2011

MIX
Papier aus verantwortungsvollen Quellen
Paper from responsible sources
FSC® C105338

Printed by Books on Demand, Germany